Robust Design of Digital Circuits on Foil

Covering both TFT technologies and the theory and practice of circuit design, this book equips engineers with the technical knowledge and hands-on skills needed to make circuits on foil with organic or metal-oxide based TFTs for applications such as flexible displays and RFID.

It provides readers with a solid theoretical background and gives an overview of current TFT technologies including device architecture, typical parameters, and a theoretical framework for comparing different logical families. Concrete, real-world design cases, such as RFID circuits and organic and metal-oxide TFT-based 8-bit microprocessors, enable readers to grasp the practical potential of these design techniques and how they can be applied.

This is an essential guide for students and professionals who need to make better transistors on foil.

Kris Myny is Senior Researcher in the Large Area Electronics Department at IMEC, Leuven. He received the 2010 IMEC Scientific Excellence Award, and the 2011–2012 IEEE SSCS Pre-doctoral Achievement Award.

Jan Genoe is Senior Principal Scientist in the Large Area Electronics Department at IMEC Leuven; Head of the Polymer and Molecular Electronics group; and a part-time Professor at the KU Leuven. He is a member of the ISSCC's Technology Directions Sub-committee.

Wim Dehaene is a Professor at the KU Leuven and the Head of the ESAT-MICAS research division. He is also a part-time Principal Scientist at IMEC, a senior member of the IEEE, and a member of the technical program committee of ESSCIRC.

Robust Design of Digital Circuits on Foil

KRIS MYNY
IMEC

JAN GENOE
IMEC

WIM DEHAENE
KU Leuven

CAMBRIDGE
UNIVERSITY PRESS

CAMBRIDGE
UNIVERSITY PRESS

University Printing House, Cambridge CB2 8BS, United Kingdom

Cambridge University Press is part of the University of Cambridge.

It furthers the University's mission by disseminating knowledge in the pursuit of
education, learning and research at the highest international levels of excellence.

www.cambridge.org
Information on this title: www.cambridge.org/9781107127012

© Cambridge University Press 2016

First published 2016

Printed in the United Kingdom by Clays, St Ives plc

A catalogue record for this publication is available from the British Library

Library of Congress Cataloguing in Publication data
Names: Myny, Kris, 1980– author. | Genoe, Jan, 1965– author. | Dehaene, Wim, author.
Title: Robust design of digital circuits on foil / Kris Myny, IMEC,
Jan Genoe, IMEC, Wim Dehaene, KU Leuven.
Description: Cambridge, United Kingdom : Cambridge University Press, 2016. |
Includes bibliographical references and index.
Identifiers: LCCN 2015047185 | ISBN 9781107127012 (hardback)
Subjects: LCSH: Logic circuits. | Flexible electronics – Materials. | Thin film transistors. |
Digital electronics. | Metal foils. | BISAC: TECHNOLOGY & ENGINEERING /
Electronics / Circuits / General.
Classification: LCC TK7868.L6 M96 2016 | DDC 621.39/5–dc23
LC record available at http://lccn.loc.gov/2015047185

ISBN 978-1-107-12701-2 Hardback

The ideal engineer is a composite. … He is not a scientist, he is not a mathematician, he is not a sociologist or a writer, but he may use the knowledge and techniques of any or all of these disciplines in solving engineering problems.

N. W. Dougherty, 1955

A design is what a designer has when time and money run out.

James Poole, chief architect for Disneyland

Contents

Contents

Preface

Guided Tour

This book is the result of several years of electronics engineering research that was performed in close cooperation between technologists and circuit design experts. The target of the research was to bring maturing thin-film transistor (TFT) technologies to a level where real, relevant circuits can be made. Relevant in this context means scientifically relevant as well as economically feasible. To enable this, types of circuit techniques were used. The advantage of the cross-level research, however, is that it is not limited to circuit techniques alone. Also technology improvements and extensions were conceived and tested. In this way dual gate TFTs were put on the map, for example.

To achieve the mission described we originally tackled several design cases. The purpose of this was to make sure that the research really produced the key enablers to enable concrete actual design. However, that is not how we present it here. Once the results were established, it seemed more logical and technically profound to present the state of the art the other way. Otherwise said: The order in which to look for things is not necessarily the best order to present those things. Therefore, the book evolved from more theoretical background information to concrete design cases.

The book starts with an introductory chapter situating TFT transistors in their application domain. It also situates organic and metal-oxide TFTs on foil next to the other TFT technologies. Organic and metal-oxide TFT are the technologies that are used in the subsequent chapters when TFT based circuits are discussed from a digital design perspective. The introductory chapter does not strive for historical completeness. It only gives a high level overview of the different TFT technologies that are available and their (potential) field of application. Details can eventually be found in the many references in the chapter.

After the introduction the book falls into two major parts. The first part, Chapters 2 to 4, discusses the necessary background for digital circuit design with TFTs on foil. The discussion first addresses the different devices that are used in Chapter 3. The different device architectures are described and typical parameters are given. These devices are further used to design logical gates in Chapter 3. Different logical families are compared to each other in terms of energy, area, robustness, and performance. For this we use a theoretical framework that is very similar to the framework that is classically used to analyze silicon CMOS gates. Chapter 4 adds the influence of technological variability on these gates to the picture. It should be mentioned that this is work in progress. Much more, mainly experimental work, to quantify TFT on foil variability is required. In general terms it can be stated that the relatively classical variability models are most likely to be a good starting point, yet further calibration with experimental data is required. Eventually the models will have to be refined.

The last part of the book contains the most important design cases that were used during the research. Chapter 5 discusses the RFID circuits that were designed, processed, and tested. In Chapter 6 the first organic, and later metal-oxide, TFT based 8-bit microprocessors are discussed. The discussion of these circuits is interesting for the circuits as such, as they are finding their way to wider application already. Of course, the design techniques in this book can be much more widely applied than for RFID or microprocessors alone.

Glancing at the Horizon

Of course, research is never complete and never ends. Each solved research question usually opens a few new ones. It is not different in this case. It is shown in this book that circuits on foil with organic or metal-oxide based TFTs are feasible. The influence of device physics related parameters has been mapped. However, it is not yet clear where the fundamental technological limitations are situated. For that more research – and perhaps a new book – is required. This future research should be centered on the following fundamental questions: The first question is which technological improvements are needed and can be realized to further improve the intrinsic quality of the devices. Can the mobility be further enhanced? How about capacitance? Can we reduce the overlap capacitance by making the devices self-aligned? At the time of this writing this is under way. Is it possible to have complementary devices? Having p and n devices of equal or at least comparable current capability would definitely further boost the performance versus energy trade-off of the TFT based circuits.

A next group of research questions is centered on modeling. There is definitely a need for more accurate, preferably device physics based, compact models for the TFTs. This includes modeling of long- and short-range variability. Long-range variability especially needs to be further analyzed as it will be a typical design issue in large area electronics. Numerical or physical models that are accurate will be a key factor to better performing circuits with a higher yield.

Last but not least there is the question of future scaling. How about next generation TFT technologies and circuits? What is the road map for these technologies? Will there be a Moore's law for circuits on foil?

The preceding questions are not answered in this book. However, the basic information to start designing circuits on foil and eventually take the research a step further is provided. If you are technologists working in circuits on foil and you want to see how this technology can be used, this book is the place to start. The same is true if you are a circuit designer and you want to figure out what the properties of TFT on foil devices are and how to put them together effectively in digital circuits.

List of Symbols and Abbreviations

Symbol	Description	Unit
ΔL	Channel length of the pinched off region	μm
ΔV	Overshoot on the output node	V
α_1, α_2	Correction factors for the channel capacitance	
β	Current factor of the I_{DS} equation	nF/Vs
γ	Proportionality or technology factor	
ε_0	Vacuum permittivity (8.854187817)	pF/m
ε_r	Relative permittivity	
$\sigma(V_T), \sigma(\mu), \sigma(I_{on})$	Standard deviation of V_T, μ, I_{on}	
ξ	Sensitivity of the back gate	
λ	channel length modulation factor	1/V
μ and $\bar{\mu}$	Mobility and average of mobility	cm²/Vs
A_{gd}, A_{gs}	Area of the gate-drain and gate-source capacitance	μm²
A_{gsc}	Area of the gate-to-drain channel and gate-to-source channel capacitance	μm²
A_μ	Area proportionality constant of μ	%μm
A_{V_M}	(Small signal) gain at V_M	
A_{V_T}	Area proportionality constant of V_T	Vμm
C	Capacitance	F
C_{ch}	Channel capacitance	F
C_{ext}, C_{int}	Extrinsic and intrinsic capacitance	F
C_g	Input gate capacitance	F
C_{gcdx}, C_{gcsx}	Gate-to-drain and gate-to-source channel capacitance of transistor x	F
C_{gd}, C_{gs}	Gate-drain and gate-source capacitance	F
C_{gdox}, C_{gsox}	Gate-drain and gate-source overlap capacitance of transistor x	F
C_{gx}	Gate capacitance of transistor x	F
C_{iref}	Capacitance of the minimum sized inverter	F
C_L	Load capacitance	F
D_x	Spacing between two transistors	μm
F	Overall effective fan-out	

Symbol	Description	Unit
FW	Finger width of source and drain contacts	μm
GND	Signal ground	
H	Electromagnetic field	A/m
I_D	Current through a diode	A
I_D, I_G, I_S	Current at the drain, gate and source of a transistor	A
I_{DS}	Drain-source current $(I_D-I_S)/2$	A
$I_{DS,lin}$, $I_{DS,sat}$	Drain-source current of a transistor operated in linear or saturation regime	A
I_{on}	On-state current of a transistor	A
I_{PD}, I_{PU}	Pull-down and pull-up current	A
J	Current density	A/cm²
L	(1) Transistor's channel length	μm
	(2) Inductance	H
L_{eff}	Effective channel length	μm
N	Chain of N inverters	
N_{opt}	Optimal number of inverter stages	
OPS	Operations per second	OPS
R	(1) Resistance	Ω
	(2) Radius reader antenna	cm
$R_{eq\,l}$, $R_{eq\,d}$	Equivalent resistor of the load and drive transistor	Ω
$R_{on,\,TD}$, $R_{on,\,TL}$	On-resistance of the drive and load transistor	Ω
R_{ref}	Resistance of the minimum sized inverter	Ω
S	Sizing factor	
S_{V_T}	Distance proportionality constant	V/μm
$V_{BG,D,G,S}$	Potential of the back-gate, drain, gate, and source node, respectively	V
V_D	Voltage across a diode	V
VDD	Supply voltage	V
V_M, V_{trip}	Trip point	V
V_T	Threshold voltage	V
$V_{T,0}$	Threshold voltage with 0 V source-back-gate bias	V
$V_{T,\,drive}$, $V_{T,\,load}$	Threshold voltage of the drive and load transistor	V
W	Transistor's channel width	μm
f	Effective fan-out	
f_{opt}	Optimum value for effective fan-out	
g_m	Transconductance	S
g_o	Output conductance	S
r	Transistor ratio	
r_o	Output resistance	Ω
t_{d0}	Intrinsic delay	s

Symbol	Description	Unit
t_{ox}	Thickness of the gate dielectric	m
t_p	Propagation delay	s
t_{pHL}, t_{pLH}	Propagation delay for a transition from high to low or low to high	s

Abbreviation	Description
2T1C	2-transistor-1-capacitor
3D	Three dimensional
4k2k	3840x2160 pixels
a-IGZO	Amorphous Indium-Gallium-Zinc-Oxide
a-Si	Amorphous Silicon
AC	Alternating current
ADC	Analog-to-digital converter
ADD	Addition
Al	Aluminum
Al_2O_3	Aluminum oxide
ALD	Atomic layer deposition
ALU	Arithmetic and logic unit
AM	Active matrix
AM LCD	Active-matrix liquid crystal display
AMOLED	Active-matrix organic light-emitting diode
Au	Gold
BCE	Back-channel etch
buf3	Buffer inverter, size is 3 times normal inverter
C-2C	Capacitor-double-capacitor architecture
CMOS	Complementary metal-oxide-semiconductor
Clk, CLK	Clock
D	Drain
D-flip-flop	Data or delay flip-flop
D2D	Die to die or inter die
DAC	Digital-to-analog converter
DC	Direct current
DC-DC	DC to DC converter
DEC	Decrement
ELA	Excimer laser annealing
EMI	Electromagnetic interference
EPC	Electronic Product Code
ES	Etch stopper
EXNOR	Exclusive NOR gate
F, FF, FS	Fast, fast-fast, and fast-slow corner

Symbol	Description	Unit
F'	Compensated fast corner	
FPGA	Field-programmable gate array	
G	Gate	
HF	High frequency	
IC	Integrated circuit	
INC	Increment	
ISO	International Organization for Standardization	
JMP	Jump	
LC	Resonant circuit, consisting of an inductor and a capacitor	
LCD	Liquid crystal display	
LD	Load	
LED	Light-emitting diode	
LSL, LSR	Logic shift left and right	
LTPS	Low-temperature polycrystalline silicon	
MC	Monte Carlo	
MEC	Maximum equal criterion	
MOSFET	Metal-oxide-semiconductor field-effect transistor	
MUX	Multiplexer	
n-TFT	N-type thin-film transistor	
NAND	Not-AND logic gate	
NOOP	No operation	
NOR	Not-OR logic gate	
OLED	Organic light-emitting diode	
opcode	Operational code	
OTFT	Organic thin-film transistor	
p-TFT	p-type thin-film transistor	
PC	Program counter	
PEDOT:PSS	Poly(3,4-ethylenedioxythiophene) poly(styrenesulfonate)	
PEN	Polyethylene naphthalate	
RC	Resistor-capacitor circuit	
RF	Radio frequency	
RFID	Radio-frequency identification	
RGB	Red Green Blue	
ROM	Read-only memory	
RR	Register select bits of the microprocessor	
RT	Room temperature	
S	Source	
S, SF, SS	Slow, slow-fast, and slow-slow corner	
S'	Compensated slow corner	

Symbol	Description	Unit
Si	Silicon	
SiN_x	Silicon nitride	
SNM	Static noise margin	
SPC	Solid phase crystallization	
SUB	Subtract	
T, TT	Typical and typical-typical corner	
T_D, T_L	Drive and load transistor	
T_D, T_S	Drive and select transistor in a 2T1C pixel scheme	
TFT	Thin-film transistor	
Ti	Titanium	
UHF	Ultra-high frequency	
VHDL	Very-high-speed integrated circuits hardware description language	
VTC	Voltage transfer curve	
WID	Within die	
WORM	Write-once-read-many times	

1 Thin-Film Transistor Technologies on the Move? From Backplane Driver to Ubiquitous Circuit Enabler?

This chapter briefly describes the product requirements for active matrix display with a focus on the technology used for the backplanes of these displays. A short overview is given of the different device technologies that are used today to implement display backplanes. Next innovation in these technologies by means of organic and oxide based devices on foil is considered. Eventually this will lead to the realization of flexible displays.

In the second part of this chapter we broaden the view. Flexible electronics foil has more applications than displays alone. Other large area electronic systems are envisaged; low-cost RFID, small controllers, and so on, can also be envisaged.

1.1 Backplanes for Active Matrix Displays

Thin-film transistors are omnipresent in our daily lives, mostly as a backplane technology for displays. The rapidly growing display industry is continuously increasing the demand for better performing thin-film transistors to be integrated in next generation display panels. The panel sizes are not limited to larger backplanes for television applications alone, but also to smaller backplanes for mobile devices such as smart phones and tablets.

One tendency in this application field is steady improvement of the displayed image quality, such that the appearance of images becomes much sharper for every new product generation. The current standard for televisions is known as *high definition* and comprises a backplane array of 1920 × 1080 pixels. The next standard is defined as *ultra-high definition, or 4k2k*, with an array size of 3840 × 2160 pixels. High-end panels in the mobile phone industry continuously squeeze in more and more pixels per inch, resulting in an ultra-sharp, non-pixelated image appearance at typical viewing distance. Current state-of-the-art smart phones exhibit a pixel density above 300 pixels per inch. The resolution of the smart phone displays increases but the size of the display remains unaltered because it is dictated by the size of the phone itself. This demands smaller pixels and better performing thin-film transistors. Another important tendency in display research is the reduction in power consumption to realize green electronics and to save battery life for handheld devices. Also in this case, better performing thin-film transistor backplanes are required to reduce the display's power consumption.

Displays in televisions and handheld devices are active-matrix (AM) displays consisting of a transistor backplane actively driving a front plane. Liquid crystal (LC) is the front plane technology that is mostly used in current televisions. An alternative front plane is based on organic light-emitting diodes, or OLEDs, recently introduced in handheld devices on the market. The advantage of OLED based displays is that the front plane directly emits light. LC based front planes actually selectively block the light from a backlight. Therefore, OLED based displays usually have better contrast and are more energy efficient. An OLED is a current driven device that emits light, whereby a change in current through the OLED results in different brightness values for the OLED. This front plane technology thus requires a high current-drive compared to LCD and is more demanding for the transistor backplane.

At the current state of the art, the substrate used for the display backplanes in production is glass. The glass size – determining the number of displays per substrate – is currently at generation 10, which has a huge physical size of 2880 × 3130 mm². Next generation applications are focusing on lightweight devices, requiring ultra-slim glass plates or plastic foil to be used as substrate. Displays fabricated on plastic substrates are very appealing for future devices, because of the possibility of realizing unbreakable, flexible, or rollable displays. This will be discussed in a later paragraph.

1.1.1   Amorphous Silicon

The most widely used thin-film transistor technology for backplanes employs amorphous silicon, or a-Si, as a semiconductor [1]. The fabrication process of an a-Si transistor backplane requires only 4–5 photo masks, and it results in n-type only devices. The standard process temperature for the semiconductor is about 250°C; the maximum temperature used in the process flow is 350°C for the SiN_x gate dielectric [2]. a-Si backplanes exhibit very good uniformity, which is due to the amorphous nature of the semiconductor with an average carrier mobility between 0.5 and 1 cm²/Vs. This mobility value is sufficient for current generation displays, but might be at the limit for next generation high-end devices where the required resolution and current drive increase. As a result of the key merits of a-Si transistor backplanes, which are technology maturity, good uniformity, and low fabrication cost, they are still predominantly used for the current and near-future LC based displays.

Despite their wide use, a-Si transistors suffer from continuous bias instabilities resulting in threshold voltage shifts over time [3]–[5]. Current LC based displays can accommodate these shifts. However, in combination with the more stringent requirements OLED based technologies pose for backplane devices, threshold shifts will become prohibitive. A threshold voltage shift introduces a variation of the source-drain current of the a-Si transistor, which directly translates into non-uniform light emission in OLED displays. Compensating for this will lead to more complex pixel circuits [6]–[9]. The need to design multiple transistors per pixel will finally limit the minimum pixel size (combined with the low charge carrier mobility) and therefore the final display resolution [10]. A more stable transistor technology than a-Si will thus be required in the near-future if the evolution in display resolution is to be sustained.

1.1.2 Low-Temperature Polycrystalline Silicon

Low-temperature polycrystalline silicon (LTPS) thin-film transistors are an alternative backplane technology to realize current and next generation AMLCD or AMOLED displays. LTPS transistors can be fabricated by multiple methods, among others excimer laser annealing (ELA) and solid phase crystallization (SPC) [11]–[14]. ELA LTPS transistors can exhibit a mobility exceeding 100 cm^2/Vs as a result of the crystallinity of the semiconductor. ELA also allows fabrication of both p-type and n-type transistors. LTPS backplanes are currently being manufactured for high-end displays given the enhanced mobility compared to a-Si transistors and the process temperature, which is still within the budget of glass substrates.

LTPS technology is also a viable candidate to fulfill next generation display requirements, such as reduced power consumption and faster switching backplanes. Moreover, the availability of a complementary (p-type and n-type) technology paves the way to integrate the peripheral display driver circuit directly on the display backplane, resulting in reduced wire fan-out [15]. Some drawbacks for this technology compared to a-Si are the higher process temperature, the device-to-device non-uniformity, and the higher cost and lower yield caused by a more complex process flow with additional mask steps [16]. This increased cost of the backplane technology is partly compensated by a reduced system cost due to the integration of the peripheral display driver circuit. The device non-uniformity stems from the presence of grain boundaries and the limited laser exposure area in one shot [13]. As a result, the pixel engine for AMOLED displays will be more complex than a traditional 2-transistor 1-capacitor (2T1C) circuit in order to compensate for V_T-mismatch in the panel area [6], [17], [18]. This is nevertheless acceptable as the footprint of the transistor can be much reduced compared to a-Si TFT, as a consequence of the better intrinsic performance. This allows integrating more complex pixel engines that can compensate for even more issues than V_T non-uniformities, for example, OLED degradation.

1.1.3 Organic Thin-Film Transistors

Research on organic thin-film transistors became very appealing as a result of the foil-compatible process temperature for these semiconductors combined with a performance similar to a-Si TFTs [19]. At the current state of the art, organic thin-film transistors are even outperforming a-Si TFTs with carrier mobility up to 10 cm^2/Vs [20]–[24]. Therefore, organic TFTs is another viable candidate for next generation display technologies, especially on flexible substrates. Moreover, the processing techniques are not limited to expensive high-vacuum coating; solution casting [20]–[23]; low-vacuum deposition [25], [26]; and even deposition at atmospheric pressure [27] are also possible. The applicability of solution-processing organic materials opens the perspectives of a new application field focusing on low-cost printed and flexible electronics. Gelinck et al. have demonstrated integrated circuits fabricated by polymer layers only, even for the contacts [28]. Also, printed transistors and circuits have been demonstrated [29]–[31]. It should be said that printed technology at the current state of the art suffers from excessively high device variability. Most organic transistors are p-type transistors. Recently, n-type TFTs have been explored to match with p-TFTs for complementary technology applications. The main disadvantages of organic thin-film transistor technologies are lack of environmental and bias stability. However, recent developments show major improvements for these instabilities [22], [24]. Finally, organic semiconductors tend to form polycrystalline layers with randomly distributed grain boundaries resulting in larger device non-uniformities compared to a-Si TFTs. The bias instabilities and non-uniformities also imply the need for more complex pixel compensating circuits for AMOLED displays [32]–[34]. This will ultimately limit the resolution of the displays that can be designed with organic TFT based backplanes.

1.1.4   Metal-Oxide Thin-Film Transistors

The field of metal-oxide thin-film transistors gained worldwide interest after the report of Nomura et al. about a single-crystalline metal-oxide semiconductor yielding great performance (carrier mobility of 80 cm^2/Vs), which truly demonstrated the potential of this type of semiconductor [35]. In the subsequent year, the same group fabricated amorphous indium-gallium-zinc-oxide (a-IGZO) TFTs on flexible substrates resulting in large mobility compared to a-Si TFTs: up to 9 cm^2/Vs [36]. Because of the potential low process temperature, the usage of conventional deposition techniques, and the promising transistor performance [1], [37], [38], metal-oxide TFTs are undoubtedly candidates for next generation display technologies and are currently already present as backplane in today's consumer products. Jeong et al. [39] reported recently on the threshold voltage uniformity of a-IGZO TFTs, exhibiting a standard deviation of 0.1 V for long-range uniformity and 0.01 V for short-range uniformity. This is much better than the 0.1 V standard deviation for short-range uniformity in ELA TFTs. This is classically attributed to the amorphous nature of the oxide in the TFTs. As a consequence, a simple 2T1C pixel engine for AMOLED displays will be sufficient. Also the line driver can be integrated [39].

Line drivers in metal-oxide TFT technology are much better candidates for integration on the backplane. The higher performance of oxide technology results in faster circuits necessary to achieve the frame rates and smaller peripheral IC size. All of this sounds like a success story; however, the downside of the medal for this semiconductor is the bias instability [37], [38]. This instability may result in more complex pixel engines, after all, which will require more area compared to simple 2T1C pixel engines. Moreover, complementary circuits are not feasible in

Table 1.1. Overview of different TFT technology options with their properties

TFT technology	Amorphous silicon	Low-temperature polycrystalline silicon	Organic	Amorphous metal-oxide
Charge carrier mobility [cm²/Vs]	0.5–1	~30–100	0.1–10	5–50
Device uniformity	Good	Poor	Poor	Good
AMOLED pixel engine	Complex	Complex	Complex	Simple / complex
Bias stability	Poor	Good	Poor	Under evaluation
Process temperature	250°C 350°C (SiN$_x$)	~500°C	RT	RT – 350°C
Semiconductor	n-type	CMOS	p-type n-type possible	n-type
Mask steps	4–5	6–9	4–5	4–5
Cost	Low	High	Low	Low

the current state of the art, since the performance of p-type oxide semiconductors is still very limited compared to that of n-TFTs. The availability of a complementary technology would be beneficial for robustness, speed, and power of the peripheral IC, which then would not be limited to a line driver only. As a final note, very recently, metal-oxide thin-film transistors and circuits have been demonstrated by means of solution-processing at room temperature, which potentially removes vacuum steps in the manufacturing flow [40].

1.1.5 Current TFT Technology Overview

Table 1.1 lists the properties of the previously discussed TFT technologies. Even though a-Si TFTs are the most widespread technology in production for present-day backplanes, their low charge carrier mobility will be insufficient to comply with next generation display standards. These future displays require low-power operation, exhibit ever-increasing panel resolutions, and will most likely introduce AMOLED as a front plane technology. AMOLED demands more and more precise current from the backplane technology. All these requirements lead to the need for better performing backplane technologies. LTPS and amorphous metal-oxide TFTs are the most promising candidates. Key advantages for LTPS backplanes are the intrinsically higher performance and the potential to integrate p-type and n-type TFTs simultaneously. This advantage allows integrating a complex peripheral circuit that reduces the overall system cost. Also amorphous metal-oxide TFTs exhibit improved charge carrier mobility compared to a-Si TFTs. Moreover, metal-oxide semiconductors can be deposited using conventional tools, enabling fast integration in current a-Si fabs. Metal-oxide backplanes can be fabricated with a larger yield compared to LTPS because of a less complex process flow and the possibility of using a simple pixel engine. All these merits for metal-oxides may lead to the use of this technology for future display panels, in spite of the availability of LTPS TFTs in today's high-end products. Finally, we should remark that a hybrid technology with n-type metal-oxide transistors and p-type organic transistors can be envisaged in the future to combine the best of both worlds.

1.1.6   Options for Flexible Displays

Unbreakable, flexible, and even rollable displays are attractive options for integration of displays in all sorts of future applications. Envisage a rollable tablet computer, for example. Rollable displays imply the necessity to replace the rigid glass substrate by a flexible substrate such as ultra-thin glass, stainless steel foil, or plastic foil. In this section, the following technological solutions resulting in flexible TFT backplanes will be briefly discussed:

- Transfer to foil,
- Direct processing on high-temperature foil,
- Direct processing on low-temperature foil.

The first option, transfer to foil, does not require low process temperatures during backplane fabrication and is therefore mostly beneficial for silicon TFT technologies [41], [42]. At the end of the fabrication flow, the stack on foil is removed by laser debonding from the temporary carrier. After the AM LCD panels, AMOLED displays and even an 8-bit microprocessor have already been demonstrated by this method [43]–[45].

The other two options require matching of the maximum process temperatures to the temperature budget of the foil. There are several substrate options available: metal foils (such as stainless steel foil) or polymer substrates. Table 1.1 suggests that organic and amorphous metal-oxide TFTs may offer proper matching of process temperature budgets to high-temperature foils and even to low-temperature foils (e.g., <180°C for polyethylene naphthalate (PEN) foil), without severe impact on the device performance [1], [19], [37], [38], [46]. Noda et al. demonstrated in 2010 the first rollable AMOLED display based on organic TFTs [47]. Finally, substrates made of paper or even banknotes are within the possibilities for organic and metal-oxide TFTs, opening perspectives for new potential application fields [48]–[51].

High-temperature foils allow the highest temperature to be maintained in the process flow. This occurs when SiN_x is processed as gate dielectric (350°C). Stainless steel foil has been explored for a-Si TFTs [52] and is considered as an option for LTPS TFTs [53]. Stainless steel foil, however, does not give a solution for fully transparent displays. For this purpose, different high-temperature foils, such as (colorless) polyimide, are being investigated for a-Si TFTs [2], [54], [55] and for LTPS TFTs by optimizing process conditions [56]. As a final note, low-temperature foils are also being investigated for a-Si TFTs [57].

1.2　　Large Area Sensors and Circuits (On Foil)

Besides displays, there are a variety of application domains that also benefit from the fast progress in TFT research driven by display applications. X-ray imagers are an example that requires a TFT backplane [1] and where the research field currently explores the option of flexible X-ray detector arrays [58]. Moreover, TFT technologies developed for flexible substrates may provide options for flexible smart systems and lower-end digital electronics.

Conventional monocrystalline silicon complementary metal-oxide-semiconductor (CMOS) technologies would easily meet the technical specifications for the applications named, such as power consumption, operational frequency, and low supply voltages. However, specifications such as low cost, large area, or flexible substrates will be difficult to obtain for monocrystalline silicon CMOS. TFT technologies, on the other hand, enter the picture when a low-cost technology per unit area is required, possibly fabricated on flexible substrates. Monocrystalline silicon CMOS technologies are currently being manufactured on 300 mm substrates targeting 450 mm as a next baseline. Thin-film transistor backplanes in contrast are fabricated on large glass plates. The size of such plates has been increased in different generations, leading to the possibility to produce larger displays. This trend is beneficial from a cost perspective, since more displays of the same size can fit on a larger substrate. A generation 8 glass substrate measures 2160×2460 mm², which may already suit a variety of large area applications. Monocrystalline silicon CMOS technologies are certainly in the lead in terms of cost per transistor, since downscaling of each technology node results in ever more transistors per unit area. TFT technologies cannot compete in achieving this extreme low cost per transistor; on the other hand, the cost per unit area to manufacture a TFT backplane is substantially lower. Moreover, solution-processing of organic and eventually metal-oxide TFTs offers a perspective to evolve toward printed low-cost electronics [1], [19].

Also in the world of ambient intelligence and the Internet of things, flexible smart systems have their place. Sensors, actuators, and cryptographic and other electronic functions can all be integrated on foil, illustrated in Figure 1.1. The integration of an AMOLED display on passports is one example of such a flexible smart system [59]. As another example, Someya et al. have demonstrated a flexible pressure sensor matrix on a large area driven by organic TFTs for artificial skin applications [60].

Integration of electronics with sensors on foil requires an efficient *analog* sensor interface. In recent years, the first analog-to-digital converters based on organic TFTs have been demonstrated [61], [62]. The digital signals from the analog-to-digital converter could be analyzed by lower-end digital electronics before storage or transmission via RF interface. A first 8-bit, flexible microprocessor realized with organic transistors that can execute basic signal analysis has been demonstrated [63]. This microprocessor is described in detail in Chapter 6.

A big driver in the field of organic and printed electronics is the application domain of radio-frequency identification (RFID) tags. When the price of such

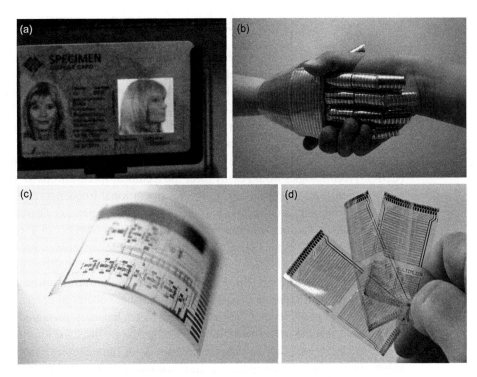

Figure 1.1   (a) Integration of an AMOLED display on passports [59]; (b) example of an artificial skin application [60]; (c) flexible, organic analog-to-digital converter [64]; and (d) 8-bit organic microprocessors on foil [63].

RFID tags can be reduced below \$0.01, the targeted application can be item-level tagging. In that case RFID on foil can eventually replace barcodes cost effectively. Organic or metal-oxide RFID tags can be used as anti-counterfeit protection, in ski passes, and in many other applications. Organic and metal-oxide thin-film transistors exhibit much less intrinsic performance compared to monocrystalline silicon CMOS circuits, for which current RFID standards are designed. A main focus in research toward low-cost, thin-film RFID tags is thus to investigate whether these silicon inspired CMOS standards can be reached with circuits in TFT technologies. Chapter 5 will discuss the design of organic and metal-oxide RFID tags.

The fact that an 8-bit microprocessor and complex analog circuits based on organic electronics have been realized implies that these technologies have reached a certain maturity. The 8-bit microprocessor contains more than 3,000 unipolar p-type, organic thin-film transistors. This level of integration thus allows the integration of basic signal processing functions. Robust logic gates are needed for such realizations. How these can be realized will be discussed in the subsequent chapters of this book.

2 Organic and Metal-Oxide Thin-Film Transistors

Organic and especially metal-oxide thin-film transistors have great potential to be introduced as a backplane technology for next generation displays or imagers, or to be used as a thin-film transistor technology on a large area for circuit and sensor integration. The discussion in this chapter focuses therefore on both technologies that have been used in this book. Primarily, four main device architectures are described. Subsequently, the operational principle of thin-film transistors is briefly explained with and without a back gate. Some typical layout rules and important parasitic capacitors in these technologies are suggested. Next, all six technologies used for the circuit realizations in the following chapters are described. Prior to the Summary, trends in circuit integration for three different circuit topics are discussed. Those three topics are display periphery, digital circuits, and analog circuits.

2.1 Device Configurations

Figure 2.1 provides a brief overview of four main transistor architectures for single-gate thin-film transistors. These architectures can be split into several categories: top-gate versus bottom-gate devices and staggered versus coplanar structures. The main difference between staggered and coplanar architectures is the location of the source-drain electrodes with respect to the transistor's channel. In a coplanar device, the source-drain electrodes are both located at the channel interface of the semiconductor. In the staggered structure, source-drain contacts are made on the opposite semiconductor interface, with the result that injected carriers at the source need to travel through the semiconductor to reach the channel. A staggered structure has a strong vertical field enhancing charge injection from the contacts. This field is shielded by the contacts in a coplanar device, which increases the injection barrier. All devices, except for the bottom-gate coplanar device, require processing on top of the semiconductor. Therefore, the semiconductor needs to be resistant to the next processing steps, when the device architecture differs from bottom-gate coplanar. Many advantages and disadvantages for these transistor architectures are extensively reviewed by Klauk for organic TFTs [19].

Park et al. elaborated on the device structures regarding metal-oxide TFTs [38]. Commonly used is the bottom-gate staggered structure, including a protective layer

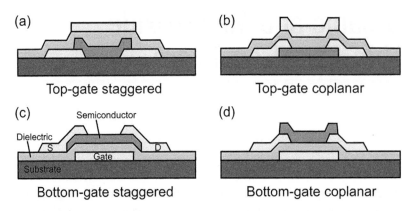

Figure 2.1   (a) Top-gate staggered, (b) top-gate coplanar, (c) bottom-gate staggered, (d) bottom-gate coplanar TFT configurations.

to isolate the back channel surface [38]. This protective layer can be added after formation of the source-drain contacts, which serve as an encapsulation layer, resulting in back channel etch (BCE) devices. Another option is to fabricate this protective layer prior to the definition of the source-drain electrodes in order to protect the back channel from further processing, resulting in etch stopper (ES) devices. ES device architectures are widely used for the fabrication of a-Si backplanes [37], [38]. As an alternative solution, Sony demonstrated in 2012 an AMOLED display based on a self-aligned, metal-oxide transistor backplane with a top-gate coplanar-like TFT [65]. The key advantage of the latter process flow is the reduction of the overlap capacitance between source-drain and gate electrodes resulting in a faster RC time to store data on each pixel.

The devices in this book have been fabricated using the bottom-gate coplanar TFT architecture, for both organic and metal-oxide semiconductors. This device structure requires for organic TFTs proper surface conditioning, both for dielectric and for source-drain contacts, in order to enhance the electrical properties of the transistor [19]. In many cases, the semiconductor has been protected for moisture by an additional encapsulation layer.

2.2 Operation Principle

2.2.1 Operation Principle of a Single-Gate Transistor

A transistor acts as an electrical switch of which the resistance depends on the voltage applied to the gate electrode. Transistors made in a classical bulk semiconductor (e.g., silicon) usually operate in inversion: source and drain diodes are in this case formed by implantation in the semiconductor of the dopant, which provides the opposite charge carrier compared to the dopant of the bulk of the semiconductor. The gate electrode is subsequently biased such that similar opposite charges are

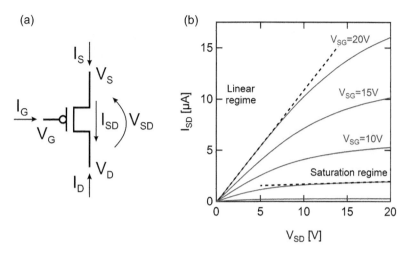

Figure 2.2   (a) Symbol of a p-type thin-film transistor and (b) output ($I_{DS} - V_{DS}$) characteristics of a typical p-type organic thin-film transistor.

obtained at its interface with the gate dielectric, effectively creating a conductive path between the source and the drain. This is called (strong) inversion.

A thin-film transistor, however, operates in accumulation. Since the semiconductor is very thin, it can be fully depleted under the gate to switch off the transistor (depletion regime). Most semiconductors used in this book are unipolar: they can only support one type of carrier. Hence, the transistor can be depleted, but inversion cannot occur. Only in special cases is inversion obtained, that is, when the gate is biased toward depletion and a source of carriers with opposite polarity is available. Such inversion operation is usually unwanted and will not be used in the present work but can be exploited in some circumstances in so-called ambipolar transistors [66], [67].

When the gate is biased toward accumulation, a conductive channel is formed between source and drain (formed as ohmic contacts to the semiconductor film). The channel is formed initially by the intrinsic conductivity of the (unintentionally) doped semiconductor, and later by the accumulated charge carriers. This occurs when the gate-source voltage is larger than the threshold voltage (V_T). A typical $I_{DS} - V_{DS}$ curve of a p-type organic TFT is shown in Figure 2.2 (b). The symbol of a p-type thin-film transistor is shown in Figure 2.2 (a). When a device is measured, we measure the current at source (I_S), drain (I_D), and gate (I_G). We typically plot the measured source-drain current I_{DS} as ($I_D - I_S$)/2, as this cancels most accurately the effect of the gate currents, and hence corresponds best to the theoretical derived I_{DS}.

The behavior of these TFTs can be described in two regimes: linear regime at low V_{DS} and saturation regime at large V_{DS}, as depicted in Figure 2.2. Figure 2.3 illustrates schematically the differences in charge carrier distribution in the channel when the transistor is operated either in linear or in saturation mode. In linear regime, a channel of charge carriers is formed between source and drain, yielding

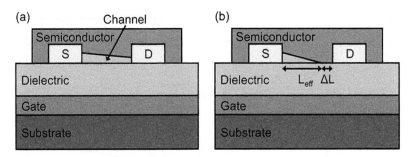

Figure 2.3   Charge carrier distribution between source and drain in (a) a linear and (b) a saturation regime.

a source-drain current that is linearly dependent on V_{DS}. The transistor acts here as a voltage-controlled resistor. The value of this resistance can be influenced by $V_{GS} - V_T$. This regime is maintained as long as V_{GS} is not too different from V_{GD}. Non-idealities such as contact resistance could be present [68], [69] introducing non-linearities in the behavior.

When the transistor is biased at larger V_{DS}, in other words, when V_{DS} approaches V_{DG}, the channel close to the drain contacts is pinched off because of the local field distribution, as illustrated in Figure 2.3 (b). As a consequence, the source-drain current becomes independent of the source-drain voltage and saturates. The transistor acts now as a voltage-dependent current source, which scales quadratically with $V_{GS} - V_T$. The transition between the linear and the saturation regime occurs at $V_{DS} = V_{GS} - V_T$. Equations 2.1 and 2.2 describe the source-drain currents for a p-type field-effect transistor in linear and saturation modes, respectively.

$$I_{DS,lin} = \mu C_{ox} \frac{W}{L} \left[(V_{GS} - V_T) V_{DS} - \frac{V_{DS}^2}{2} \right], \text{ when } V_{DS} < V_{GS} - V_T \qquad (2.1)$$

$$I_{DS,sat} = \frac{1}{2} \mu C_{ox} \frac{W}{L} (V_{GS} - V_T)^2, \text{ when } V_{DS} > V_{GS} - V_T \qquad (2.2)$$

Figure 2.2 also shows that the current does not fully saturate in saturation regime, but increases by increasing V_{DS}. Indeed, the ideal carrier profile in saturation should be independent of the drain voltage, but in practice, there is still an influence of the drain voltage on the carrier profile in the channel and hence the obtained current. This influence of the drain voltage depends on the actual transistor architecture (e.g., staggered versus coplanar, BCE or ES, dual gate or single gate, dielectric thickness versus semiconductor thickness, and channel length, and so on). The detailed modeling of the saturation behavior goes beyond the scope of this dissertation. Most of the charge carrier profile changes mentioned can be represented by a change in effective channel length (L_{eff}), illustrated in Figure 2.3 (b) and mathematically represented by Eq. (2.3). When the drain-source field increases, the pinch off region at the drain contact also increases, resulting in a decrease of L_{eff}.

Equation 2.4 describes the drain-source current for the saturation regime, taking the effective channel length into account. It shows that the drain-source current increases as a consequence of the smaller channel length. This effect is less present for larger channel lengths.

$$L_{eff} = L - \Delta L \tag{2.3}$$

$$I_{DS,sat} = \frac{1}{2}\mu C_{ox} \frac{W}{L_{eff}}\left(V_{GS} - V_T\right)^2 \tag{2.4}$$

To model this effect as a function of the drain voltage, the channel length modulation parameter λ has been introduced; it is typically inverse proportional to the channel length. Equation 2.2 including λ becomes

$$I_{DS,sat} = \frac{1}{2}\mu C_{ox} \frac{W}{L}\left(V_{GS} - V_T\right)^2\left(1 + \lambda V_{DS}\right), \text{ when } V_{DS} > V_{GS} - V_T \tag{2.5}$$

These equations describe the simplest model for TFTs that we have used for circuit analysis. Furthermore, circuit simulation for organic and metal-oxide thin-film transistors has been performed by using simple low-level spice models and even a more advanced level-61 a-Si TFT model. The parameters for these spice models are subsequently extracted by curve fitting. Moreover, compact analytical models describing organic or metal-oxide TFTs are widely studied in the literature for potential integration into a spice environment with better accuracy [70]–[80].

2.2.2 Technology Options for Multiple Threshold Voltages

In the next chapter, we discuss different logic gate implementations. The availability of a second threshold voltage in a unipolar technology is beneficial for the characteristics of these logic gates. For Si CMOS transistors, the threshold voltage can be accurately set by the amount of doping applied by ion implantation [81], [82]. Technologically, it is not straightforward to obtain multiple threshold voltages for organic and metal-oxide thin-film transistors. Several published options are discussed in the next part.

Nausieda et al. demonstrated dual-V_T inverters and ring oscillators by the formation of two different gate metals, a high-V_T and low-V_T gate metal [83]. In 2012, Han et al. tuned the threshold voltage of organic TFTs by doping with gold nanoparticles. By this method, unipolar dual-V_T inverters are demonstrated [84]. Another option suggested in the literature to control the threshold voltage is programming a floating gate, demonstrated by Yokota et al. for complementary circuits [85]. Alternatively, treatments to the dielectric prior to the deposition of the organic semiconductor [86], [87] or surface treatments of the source-drain contacts [88] could be a viable route to adapt the threshold voltage.

Besides previous options, the multiple threshold-voltage implementation in this work is based on a dual-gate transistor [89]–[92]. The schematic cross section of this device is shown in Figure 2.4. The transistor comprises two gates, namely,

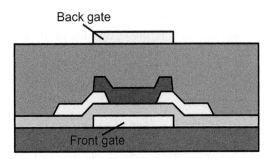

Figure 2.4   Cross section of a dual-gate thin-film transistor.

a front gate and a back gate. Both gates can deplete or accumulate charges in the semiconductor by their electrostatic field. Depending on the thickness and the roughness of the top interface of the semiconductor, one or two channels can be present [92], [93]. In this book, we assume that only one channel is available. This is typically the case when the roughness of the upper semiconductor interface hampers the transport (as in evaporated pentacene devices) or when the semiconductor is too thin (e.g., when only a few-nanometer a-IGZO is deposited). The basic operation principle is explained for p-type TFTs.

Biasing the back-gate voltage (V_{BG}) positively with respect to the source will deplete the accumulated charges in the channel. In order to switch on the channel, a smaller (or more negative) bias voltage to the front gate (V_{GS}) is required. In fact, the threshold voltage has been shifted to smaller or more negative values. When $V_{BG,S}$ is negatively biased, on the contrary, charges are being accumulated in the channel. As a consequence, V_{GS} needs to be larger in order to deplete the channel fully. This leads to a positive-shifted threshold voltage. Section 2.4.2 discusses typical characteristics and its parameters for the devices used in this work.

2.3   Typical Layout Rules in the Technologies Used in This Book

This section describes some important layout rules for technologies used in this book. A cross section of all layers to be patterned is shown in Figure 2.5 (a): the gate, dielectric, source-drain contacts, and the semiconductor. Figure 2.5 (b) depicts a typical top view of an interdigitated layout. The latter layout is used for transistors with large width/length ratios (such as a $W/L = 140/5$ transistor) in order to reduce the final device area. Generally, a minimum design rule of 5 μm is chosen to allow for some misalignment and to take the dimensional stability of foils during processing into account.

The gate should extend the semiconducting island; otherwise a parasitic conductive path between source and drain contacts can be present for normally-on technologies. This results in an increase of the off-current. A typical minimal channel length (L) for most of the designs is chosen to be 5 μm, except for the technology

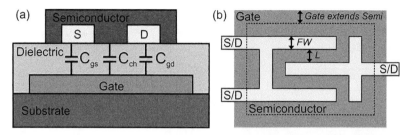

Figure 2.5 (a) Device cross section of a typical organic TFT in our technologies and (b) top view of a typical interdigitated layout illustrating some layout rules.

described in Section 2.4.3. In that section, the aim is to improve on the circuit speed by decreasing channel length. The minimal channel length used for those devices is 2 μm.

The integrated circuits realized in the technologies discussed in the next section have not been fabricated using a *self-aligned gate* process flow. This implies that (major) parts of the source and drain contacts are overlapping with the gate metal. This forms large parasitic overlap capacitances (C_{gs} and C_{gd}) between gate and source or drain, respectively. The parameter FW in Figure 2.5 (b) is defined as the critical finger width of the source and drain contacts; 5 μm is chosen as a standard rule. The impact of decreasing FW to 2 μm is studied for the technology of Section 2.4.3, which is used to improve the circuit's operational frequency with reduced parasitic overlap capacitances.

The minimum dimension for via holes that connect the source-drain to the gate metallization layer is taken at least 5 μm at each side. For the same rule, both the source-drain and gate metallization layers should extend the via connection at least 5 μm at each side.

2.4 Technologies Used in This Book

2.4.1 Organic p-Type Technology of Polymer Vision

The organic electronics technology in this section has been developed by Polymer Vision for commercialization in rollable active matrix displays and is described elsewhere [94], [95]. Figure 2.6 (a) shows the cross section of the layer stack. These transistors are realized on a 150-mm-diameter, 25-μm-thick flexible substrate that is laminated on a rigid carrier. The gate layer, Au, serves as the bottom gate in this coplanar transistor configuration. A 350-nm polymer gate dielectric is processed from solution. Via connections between the gate and source-drain level can be defined by UV exposure of the photopatternable gate dielectric. Au source and drain layers are subsequently defined, leading to typical channel lengths of 5 μm. The gate and source-drain metallization layers are used to realize all interconnects. A precursor pentacene is spin-coated on the wafer and converted into pentacene,

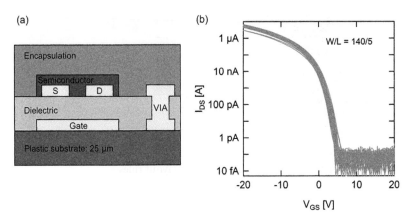

Figure 2.6   (a) Cross section of a single-gate pentacene thin-film transistor and (b) 16 transfer curves visualizing the local variation, obtained at $V_{DS} = -20$ V for a transistor with channel length of 5 μm and width of 140 μm.

which is the semiconductor [20]. As a final step, an organic insulating layer is deposited as an encapsulation layer, because pentacene is prone to degradation in ambient [96]. The transistors exhibit an average saturation mobility of 0.15 cm²/Vs, yielding normally-on characteristics with an onset voltage around 4 V; see Figure 2.6 (b). The smallest transistor in our digital circuits has a channel width of 140 μm. The extracted standard deviation for the threshold voltage is 0.35 V for a 140/5 transistor. Further mismatch analysis has been evaluated in more detail in Chapter 4. This technology is used for circuits in Chapter 5, for the first and second generation of transponder chips (Sections 5.2 and 5.3) [97].

2.4.2   Organic p-Type Dual-Gate Technology of Polymer Vision

The technology cross section in this section is the same as Figure 2.6, besides an additional metallization layer to form the back gate or V_T-control gate. The top insulator is approximately 1.4 μm thick and acts as the gate dielectric for the back gate [89]. The cross section of this technology is depicted in Figure 2.7. Via connections between each metallization layer are possible. As a side note, this third metallization layer can be used as an additional interconnect layer that can lead to smaller chip area.

The effect of applying voltage biases to the back gate or V_T-control gate is depicted in Figure 2.8. By varying the back-gate bias between −30 V and +30 V, the transistor's threshold voltage can be controlled, leading from depletion-mode to more enhancement-mode behavior, in agreement with earlier publications [89], [90].

Figure 2.9 shows the extracted threshold and onset voltage as a function of the source-back-gate voltage V_{SBG}. Please note this is the inverse polarity of the V_{BGS} voltage used in Figure 2.8. V_{on} is defined as the onset voltage where a current level of 1 pA is reached in the transfer curve. A linear relationship between threshold voltage and back-gate voltage with respect to V_{SBG} is observed.

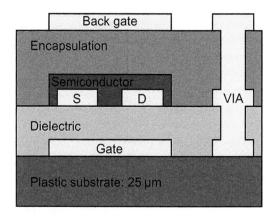

Figure 2.7 Transistor stack of a dual-gate organic thin-film transistor.

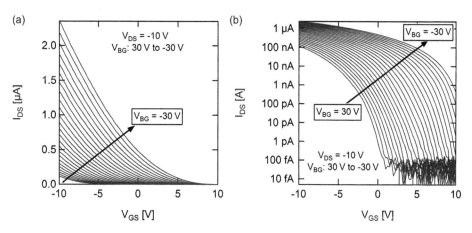

Figure 2.8 Transfer characteristics of the dual-gate transistor in saturation plotted on (a) a linear and (b) a logarithmic scale, when varying V_{BG} between -30 V and $+30$ V.

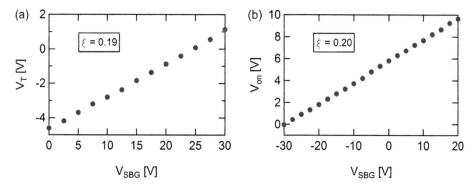

Figure 2.9 (a) V_T and (b) Von plotted as a function of the source-to-back-gate voltage V_{SBG}.

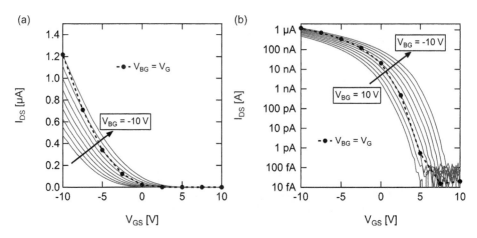

Figure 2.10   Transfer characteristics of the dual-gate transistor in saturation plotted on (a) a linear and (b) a logarithmic scale, when varying V_{BG} between −10 V and +10 V. The dots correspond to the measurement points when V_{BG} is equal to V_G. The dashed line connects these dots as a guide for the eye.

The threshold voltage can be shifted linearly by ξ multiplied by the source-to-back-gate voltage, described in Eq. (2.6) [61]. $V_{T,0}$ is defined as the threshold voltage with 0 V source-back-gate bias.

$$V_T = V_{T,0} - \xi V_{BGS} \tag{2.6}$$

The sensitivity of the back gate is defined as ξ. The theoretical value for ξ is the ratio between top and bottom gate capacitance, being, respectively, 2 and 10 nF/cm² according to [89], as shown in Eq. (2.6).

$$\xi_{theoretical} = \frac{C_{top}}{C_{bottom}} = \frac{2nF / cm^2}{10nF / cm^2} = 0.2 \tag{2.7}$$

The source-drain current equation in saturation can now be rewritten as follows [61]:

$$I_{DS,sat} = \frac{1}{2}\mu C_{ox}\frac{W}{L}\left(V_{GS} + \xi V_{BGS} - V_{T,0}\right)^2 \tag{2.8}$$

The transconductance is derived in Eq. (2.9). The back-gate voltage can increase or decrease the transconductance by a factor of $\mu C_{ox}\dfrac{W}{L}\xi V_{BGS}$.

$$g_m = \mu C_{ox}\frac{W}{L}\left(V_{GS} + \xi V_{BGS} - V_{T,0}\right) \tag{2.9}$$

Figure 2.10 illustrates a special case of the dual-gate transistor whereby both gates are connected. The dashed line shows the corresponding transistor curve. It suggests improved characteristics, such as subthreshold swing, but also an increased g_m.

The source-drain current equation can be derived for this situation from Eq. (2.8), whereby V_{BGS} is equal to V_{GS}.

$$I_{DS,sat} = \frac{1}{2}\mu C_{ox}\frac{W}{L}\left((1+\xi)V_{GS}-V_{T,0}\right)^2 \tag{2.10}$$

Isolating $(1+\xi)$ from Eq. (2.10) and calculating the transconductance lead to the following equations:

$$I_{DS,sat} = (1+\xi)^2\frac{1}{2}\mu C_{ox}\frac{W}{L}\left(V_{GS}-\frac{V_{T,0}}{1+\xi}\right)^2 \tag{2.11}$$

$$g_m = (1+\xi)^2\mu C_{ox}\frac{W}{L}\left(V_{GS}-\frac{V_{T,0}}{1+\xi}\right) \tag{2.12}$$

The threshold voltage cannot be controlled anymore when both gates are connected. On the other hand, the source-drain current and the transconductance are increased by the term $(1+\xi)^2$. In our specific case, it introduces an increase of 44 percent for the transconductance, while the extra parasitic gate capacitance caused by the back gate is only 20 percent. This is a favorable effect for the transistor's speed performance. The dual-gate transistor with gate-to-back-gate connection has been used by Marien et al. to increase the analog gain of a differential amplifier [61].

The availability of multiple threshold voltages is beneficial to increase the robustness of logic gates, as will be discussed in Chapter 3. As a consequence, implementing an additional gate to control the threshold voltage of each individual transistor could be one of the solutions. Chapter 3 explores, therefore, various implementations of logic gates using the back-gate technique. One implementation uses the gate-to-back-gate connected transistor as a load transistor. The integration of these logic gates into larger circuits is discussed in Chapters 5 and 6.

2.4.3 Pentacene (p-Type) Thin-Film Transistors on Al₂O₃ as Gate Dielectric

The cross section of this technology is shown in Figure 2.11. Vapor-deposited pentacene [26] exhibits a charge carrier mobility of 0.5 cm²/Vs in our bottom-contact devices. In this book, 30 nm pentacene has been evaporated in an ultra-high vacuum chamber at 0.25 Å/s, whereby the substrate was kept at 68 °C. The isolation of the semiconductor film for each individual transistor is achieved by an integrated shadow mask [98], as illustrated in Figure 2.11. This results in a reliable isolation of the semiconductor area and off-currents below 10 pA reaching the noise floors of our measurement setup. As gate dielectric, we use sputtered or atomic-layer-deposited (ALD) Al_2O_3, with a specific capacitance of 70 nF/cm², treated by a self-assembled monolayer of trichloro(phenethyl)silane to enhance the pentacene film growth. That allows for channel length downscaling, in our case to 2 μm, while maintaining a reasonable output resistance in saturation of r_o = 9.75 MΩ at 0 V gate-source voltage. All devices and circuits are processed on a 25-μm-thick polyethylene naphthalate (PEN) foil, which is laminated on a 150-mm carrier substrate during processing and delaminated after completion of the process flow.

Typical transfer and output curves of transistors fabricated in this technology with L = 5 μm and 2 μm are depicted in Figures 2.12 and 2.13. The transistors

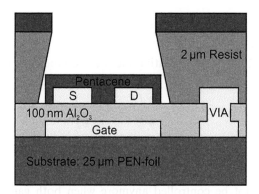

Figure 2.11   Cross section of the pentacene thin-film transistor process on foil with Al_2O_3 as gate dielectric and an integrated shadow mask to pattern the semiconductor.

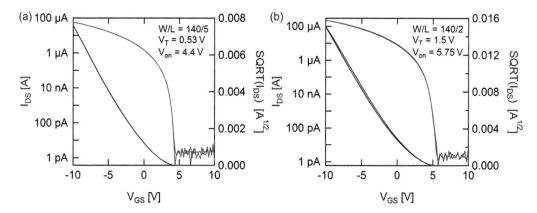

Figure 2.12   Typical measured transfer curves of transistors with (a) W/L = 140/5 and (b) W/L = 140/2. The mobility exceeds 0.5 cm²/Vs; the off-current is below 10 pA.

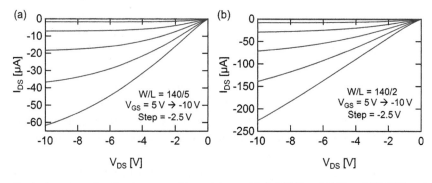

Figure 2.13   Typical measured output curves of transistors with (a) W/L = 140/5 and (b) W/L = 140/2.

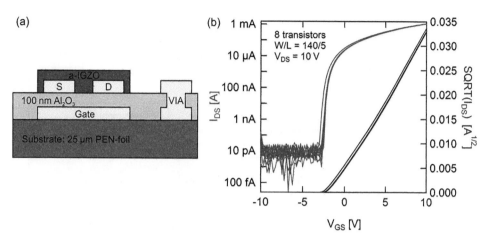

Figure 2.14   (a) Cross section of the a-IGZO semiconductor-based thin-film transistor process on foil having Al_2O_3 as gate dielectric; (b) eight transfer curves measured in a saturation regime.

are normally-on and their charge carrier (hole) mobility exceeds 0.5 cm^2/Vs. The output resistance in saturation at gate-source voltage of 0 V is 207 ± 8 MΩ for W/L = 140/5 μm and 9.8 ± 0.2 MΩ for W/L = 140/2 μm.

The benefits of this technology are further explored in Chapter 5. It allows one to design with smaller overlap capacitance and to downscale the transistor channel length, within the boundaries achievable by existing high-throughput tools (e.g., steppers used in backplane manufacturing). Specifically, we investigated the impact on circuit performance as a consequence of channel length downscaling between 20 μm and 2 μm. We also reduced the parasitic gate-source and gate-drain overlap capacitances by decreasing the width of the finger-shaped source and drain contacts from 5 μm to 2 μm, as illustrated in Figure 2.5.

2.4.4   a-IGZO (n-Type) Technology on Al_2O_3 as Gate Dielectric

This technology comprises metal-oxide TFTs that are integrated into circuits. The metal-oxide semiconductor used here is sputtered amorphous-IGZO (a-IGZO). A cross section of the a-IGZO TFT technology is shown in Figure 2.14 (a), as published by Tripathi et al. [99]. After GIZO deposition a short anneal step of 15 min at 150 °C in air was performed. The gate dielectric is atomic layer deposition (ALD) of Al_2O_3 with a specific capacitance of 70 nF/cm^2. The resulting transistors yield mobilities around 9 cm^2/Vs and exhibit an excellent local uniformity, as shown in Figure 2.14 (b).

This technology is explored in Chapter 5 as an alternative solution to increase the circuit's performance regarding speed by a larger charge carrier mobility. This is necessary to comply with the data rate specifications of RFID standards. The excellent local uniformity allowed integrating these TFTs in RFID transponder chips.

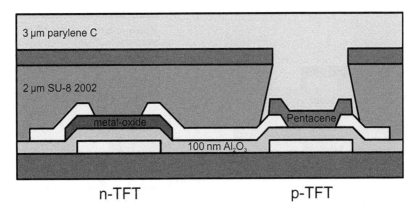

Figure 2.15   Cross section of the hybrid complementary technology having Al_2O_3 as gate dielectric.

2.4.5   Hybrid Complementary Organic/Metal-Oxide Technology

A complementary technology combines p-type and n-type transistors on the same substrate, enabling the realization of complementary circuits (similar to silicon CMOS). Complementary logic gates have many advantages compared to unipolar logic gates, which will be discussed in more detail in Chapter 3. These merits drive the research field of thin-film transistors for circuit applications toward complementary technologies. There are a variety of integration schemes possible to combine p-TFTs and n-TFTs in a reliable process flow. We discussed in Chapter 1 that organic technologies can have both p-type and n-type TFTs available while currently in the field of metal-oxide transistors, the n-TFT is still outperforming the p-TFT. Therefore, next to a fully organic complementary technology, there is also the option to combine a metal-oxide n-TFT to an organic p-TFT in a hybrid complementary process flow.

The latter process flow is used to realize RFID transponder chips in Chapter 5. It combines a 250 °C solution-processed n-type metal-oxide TFT with a typical charge carrier mobility of 2 cm²/Vs and an evaporated pentacene p-type TFT, yielding a mobility of up to 0.3 cm²/Vs after the completion of the process flow. This hybrid process builds upon a gate first stack, both for the n-type and for the p-type transistor (see the cross section shown in Figure 2.15). ALD Al_2O_3 is used as gate dielectric. The metal-oxide semiconductor is a 10-nm-thick indium-based layer (iXsenic S, developed by Evonik Industries, www.evonik.com), processed and annealed prior to the deposition of a bilayer of 2 nm Ti and 30 nm Au source-drain contacts. The n-TFT is subsequently covered by a 2-μm-thick SU-8 2002 (MicroChem Corp) that protects the n-TFT from further processing and acts as an integrated shadow mask to isolate the organic p-TFTs (similarly to Section 2.4.3). Pentacene has been evaporated in an ultra-high vacuum chamber. As a consequence of the process order, the n-TFT is a bottom-gate staggered device, and the p-TFT is a bottom-gate coplanar device. Finally, the hybrid complementary technology is capped with an encapsulation layer consisting of 3 μm parylene C. More process details can be found in [100], [101].

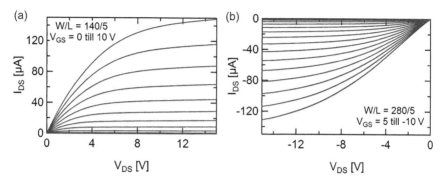

Figure 2.16 Output characteristics of (a) solution-processed metal-oxide TFTs with a W/L of 140/5 and (b) evaporated pentacene TFTs with W/L of 280/5 after finalizing the full integration flow.

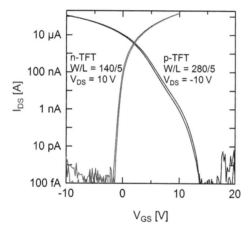

Figure 2.17 Typical measured transfer curves of p-TFTs and n-TFTs biased in a saturation regime, after finalizing the full integration flow.

Output and transfer characteristics of both TFTs are depicted in Figures 2.16 and 2.17. Because of the differences in charge carrier mobility and threshold voltages, we found the optimal p:n ratio for logic gates was 2:1 for static performance. The RFID transponder circuits discussed in Chapter 5 based on this technology are fabricated on a rigid carrier as a result of the relatively high process temperature of the metal-oxide semiconductor. Rockelé et al. have already demonstrated the possibility of using high-temperature foils enabling the integration of this hybrid complementary technology on polyimide foil [102].

2.4.6 Hybrid Complementary Organic/Metal-Oxide Technology on PEN-Foil

The hybrid complementary process flow of this section does not differ much from previously discussed technology; see Section 2.4.5. The main difference is the integration of a different solution-processed n-type material, which is a high-performance

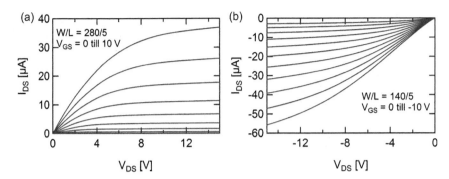

Figure 2.18   Output characteristics of (a) solution-processed metal-oxide TFTs with a W/L of 280/5 and (b) evaporated pentacene TFTs with W/L of 140/5 after finalizing the full integration flow.

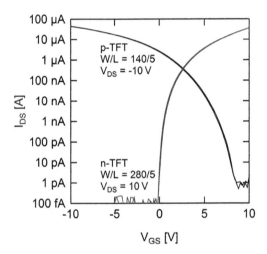

Figure 2.19   Typical measured transfer curves of p-TFTs and n-TFTs biased in a saturation regime, after finalizing the full integration flow.

ZnO-based metal-oxide. The semiconductor is applied by spin-coating and post-annealed at a temperature of only 160 °C. This low temperature enables direct integration of the transistor stack on PEN-foil. The n-TFT exhibits a charge carrier mobility after finalizing of the full integration flow around 0.25 cm²/Vs. The p-type has a mobility of approximately 0.2 cm²/Vs. Typical I_{DS}–V_{GS} characteristics are shown in Figure 2.18; the corresponding transfer curves are depicted in Figure 2.19. Because of the differences in charge carrier mobility and threshold voltages, we found the optimal p:n ratio for logic gates was 1:2 in terms of static performance. This complementary process has been used to evaluate load dependency of inverter stages in Chapter 3. Moreover, we have investigated channel length downscaling and reducing parasitic overlap capacitances to improve circuit's performance targeting RFID transponder chips in Chapter 5.

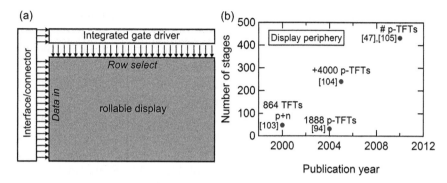

Figure 2.20 (a) Schematic representation of a rollable display with integrated gate driver; (b) progress of the periphery complexity for organic TFTs, expressed as the number of stages that can be driven and corresponding number of TFTs (in the figure).

2.5 Trends in Circuit Integration

In this section, we discuss trends observed in the research field of TFT circuits, mostly focused on organic, but also metal-oxide TFTs. Organic TFTs may find application in microelectronic circuits that can be divided into three categories. The first category is defined as circuit blocks needed for the display periphery aiming for flexible and even rollable displays, or for display panels with smaller bezels. The next category is digital circuits, such as a transponder chip for an identification tag, where Boolean information from a memory is read out and prepared for radio frequency (RF) transmission to a base station. The third type of circuits are so-called analog circuits, typically used for conditioning, filtering, and amplifying analog signals, or for conversion of analog signals into a stream of Boolean digits.

2.5.1 Display Periphery

The previous chapter discussed extensively the display industry as the main driver for organic and metal-oxide thin-film transistor technologies. One potential application that is envisioned in this industry is flexible and rollable displays. In order to achieve display with a good rollability, there is a need to replace the standard rigid silicon driver ICs by an integrated gate driver. This implies that the gate driver is fabricated in the same technology as the display's backplane. A possible example is schematically represented in Figure 2.20 (a). All signals required from the driver electronics in order to display images or movies are routed to one side for a rollable display.

Figure 2.20 (b) displays the evolution of display periphery for organic TFT backplanes driving electrophoretic and/or AMOLED displays. The gate driver needs to provide signals that select a single row, at least for basic driving schemes. An example could simply be a running 0 or 1, depending on the technology. Crone et al. had demonstrated already in 2000 the first shift register based on an organic complementary technology [103]. This shift register comprises 864 TFTs driving 48 stages. Four years later, Gelinck et al. integrated 1888 p-TFTs into a 32-stage register [94]. This is a significant increase in the number of TFTs driving fewer stages that can be explained by the choice of logic gates due to the technology limitations. This shift register has been realized in a unipolar (p-type) technology designed with logic gates using level shifters to increase operational frequency. In the subsequent year, van Lieshout et al. integrated +4000 p-TFTs into a 240-stage shift register [104]. Noda et al. demonstrated the world's first AMOLED display that could be rolled around a cylinder with a radius of 4 mm [47], [105]. The gate driver has been integrated into the display with a size of $432 \times 240 \times RGB$. No data are available on the number of p-TFTs. Although the first publication was based on a complementary technology, the next subsequent periphery circuits integrated unipolar TFTs into the driver circuit.

This integration trend for display periphery circuits has been completely opposite that of the silicon CMOS industry. An organic or metal-oxide backplane technology comprises in many cases only p-TFTs or n-TFTs. Integrating a complementary technology for the display periphery only will make the backplane process flow more complex. This will have an impact on cost and yield of the final backplane. On the other hand, display backplanes might have a third metallization layer available, functioning as a pixel electrode. Therefore, dual-gate transistors can be available that have beneficial influences when integrated into circuits in terms of robustness or operating speed. Dual-gate logic, as we will discuss in more detail in the next chapters, could pave the way for next generation rollable displays with integrated gate drivers.

2.5.2   Digital Logic

Shift registers in display periphery also contain digital logic; however, these circuits mostly consist of one basic circuit block repeated over multiple stages. In this section, the discussion will focus mainly on complex digital designs. The main driver in this field is low-cost, flexible RFID tags. A comprehensive discussion of this topic can be found in Chapter 5. Three trends for digital circuits will be discussed in this section: increasing circuit complexity, increasing operating speed, and decreasing supply voltage.

The integration density for digital logic is strongly determined by the robustness of the logic gates and the maturity of the technology. Figure 2.21 illustrates the trend in increasing circuit complexity for different technologies, being unipolar, dual-gate unipolar, and complementary. Cantatore et al. demonstrated in 2006 the first 64-bit capacitively coupled RFID system, read out at 125 kHz [106], [107]. This design employs 1938 p-TFTs. In the same technology, 8-bit RFID tags are read out at 13.56 MHz. Ullmann et al. integrated a 64-bit tag, operating at a base carrier frequency of 13.56 MHz [108]. In 2008, we demonstrated a 64-bit RFID tag, also operating at 13.56 MHz [97], [109]. The first increase in complexity of the transponder chip, though not in effective number of transistors compared to

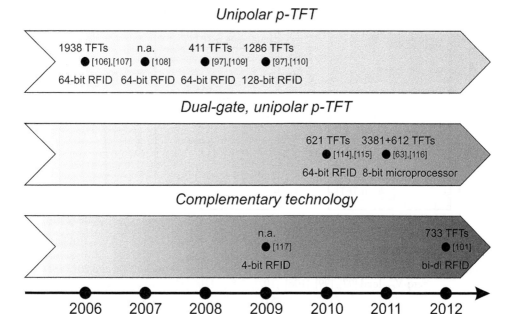

Figure 2.21   Digital design complexity (excluding repetitive designs) versus publication year for unipolar p-TFT, dual-gate unipolar p-TFT, and complementary technology.

[106], [107], was realized in 2009. The number of bits that could be read out had been increased to 128, the data were encoded by Manchester encoding, and a basic ALOHA anti-collision protocol was present [97], [110].

A next step in robustness of logic gates can be obtained by introducing a second threshold voltage in the unipolar technology, such that the load and drive TFT have different threshold voltage. More details can be found in the next chapter. Dual-gate TFTs can fulfill the requirement of two threshold voltages for integrated circuits [111]–[115]. In 2010, we studied integrated digital logic based on dual-gate organic TFTs in a 64-bit RFID transponder chip comprising 621 dual-gate p-TFTs [114], [115]. One year later, the first organic microprocessor was published, based on dual-gate p-TFTs [63], [116]. This microprocessor has been realized as two foils, an arithmetic-and-logic-unit foil and an instruction generator foil like a running averager. These foils comprise, respectively, 3381 and 612 dual-gate p-TFTs.

The best option for circuit robustness is the availability of a complementary technology. Blache in 2009 developed a 4-bit organic complementary RFID tag based on n-type and p-type organic field-effect transistors, read out at 13.56 MHz [117]. Three years later, the first bi-directional RFID tag was realized in a hybrid complementary, organic/metal-oxide technology [101].

The trend to increase the operational speed of digital circuits is shown in Figure 2.22 for unipolar logic, dual-gate unipolar logic, and complementary logic. The focus on this trend is based on further analysis of the previously discussed organic and metal-oxide RFID transponder chips. To comply with Si-based RFID

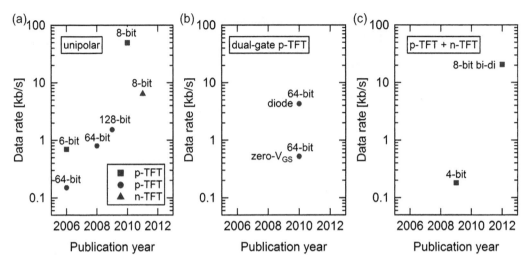

Figure 2.22   Data rates published for RFID transponder chips between 2006 and 2012 for (a) unipolar zero-V_{GS}-load [97], [99], [106], [107], [109], [110], [118]; (b) dual-gate p-TFT [114], [115]; and (c) complementary technologies [117], [101].

standards for item-level tagging, RFID transponder chips need to be able to operate at approximately 52 kbit/s. As a consequence, besides the increase of chip complexity, data rate improvements are required.

The data rates for 64- and 128-bit unipolar p-type transponder chips increased between 2006 and 2009 [97], [106], [107], [109], [110]. The circuit architecture and complexity play an important role in the circuit's speed. The 64-bit RFID transponder chip published by Cantatore et al. requires a VDD of 30 V [106], [107]; in the same work a 6-bit transponder chip was also realized. The latter chip is less complex in design architecture and employs fewer transistors. The required supply voltage is 25 V and the chip has a faster data rate compared to the 64-bit version. The 64-bit transponder demonstrated in 2008 [97], [109] implements almost fivefold fewer transistors compared to [106], [107]. The reported supply voltage is as low as 14 V. In the subsequent year, a 128-bit transponder chip with increased complexity was demonstrated, requiring a supply voltage of 24 V [97], [110]. This increase in supply voltage was not only due to increased complexity, but also to a more positive onset voltage. An increase in charge carrier mobility directly translates into faster data rates. This was demonstrated in 2011 by Tripathi et al. [99], who integrated sputtered a-IGZO as semiconductor with a larger mobility, yielding data rates of 6.4 kbit/s for an 8-bit transponder chip. The supply voltage for this chip is 2 V, suggesting a low onset voltage for the unipolar n-TFTs in this technology.

A limiting factor for speed are the large overlap capacitances C_{gs} and C_{gd}, as illustrated in Figure 2.5, because none of these circuit publications so far has been realized using a self-aligned gate technology. These parasitic overlap capacitances play an important role in the circuit's performance [19], [118]. Myny et al. reported different layouts of 8-bit transponder chips realized in the same technology on the same wafer. One explored layout difference is the reduction of the source-drain finger width *FW* (Figure 2.5) from 5 to 2 µm [118]. Channel length downscaling is another method to increase the data rates when contact resistance remains sufficiently low and the patterning technology allows for downscaling of critical dimensions [19], [118]. As a consequence, in 2010, we published the first 8-bit transponder chip complying with the silicon-based standards [118]. The realized chip with a channel length of 2 µm and a decreased parasitic overlap capacitance yields a data rate of approximately 50 kbit/s.

Another method to improve on data rates is to switch from zero-V_{GS}-load logic to diode-load logic. In 2010, data rates of 64-bit transponder chips were compared for diode-load and zero-V_{GS}-load logic in the dual-gate technology. Because of the increased circuit robustness by the addition of an extra gate, diode-load logic circuits could also be realized. Diode-load logic leads to increased data rates of the 64-bit transponder chip [114], [115].

Complementary technologies can also yield faster data rates. Blache et al. demonstrated in 2009 the first 4-bit transponder chip in an organic complementary technology [117]. The minimum supply voltage for this technology was 5 V, resulting in a data rate of 61 bit/s. At 20 V VDD, this data rate could be increased to 162 bit/s. We demonstrated in 2012 a bi-directional RFID chip realized in a hybrid complementary, organic/metal-oxide technology. This bi-directional RFID chip comprises an 8-bit code generator. The data rate of this code generator yields 20.6 kbit/s at a supply voltage of 10 V. The minimum VDD for these chips is as low as 3.75 V while still maintaining a data rate of 6.3 kbit/s [101].

The supply voltage can be decreased when the robustness of the logic gates increases, from unipolar to dual-gate unipolar and optimally to complementary logic, as discussed in previous paragraphs. Another method to reduce the circuit's supply voltage is to decrease severely the dielectric thickness. Klauk et al. reported in 2007 an organic complementary technology on very thin dielectrics, consisting of a self-assembled monolayer on ultrathin Al_2O_3 [119]. The total dielectric thickness is 5-6 nm only, yielding a complementary inverter operating at 1.5 V VDD. Several different digital circuits have been realized in this technology, all operating at low supply voltage [120]–[122]. In 2010, Ishida et al. realized a stretchable electromagnetic interference (EMI) flexible sheet consisting of an organic complementary 3-to-8 decoder at 2 V VDD [120]. The low operating voltage of this circuit allows direct signal transmission between the organic CMOS technology and the 0.18-µm CMOS chip, which is a key merit of this technology. Another circuit demonstration in this technology is a user customized logic paper based on sea-of-transmission-gates [121]. As alternative option for these organic CMOS TFTs, analog circuits have been demonstrated [62], [123].

2.5.3 Analog Circuits

Analog circuit design requires a certain maturity of the technology, which is the main reason why complex analog circuits only recently entered the picture for organic (and metal-oxide) thin-film transistors. This section will discuss DC-DC converters, digital-to-analog, and analog-to-digital converters. The trends for these circuits are similar to those for digital logic: a lower supply voltage, an increased accuracy and complexity of the converters, and faster performing circuits.

A DC-DC converter can be useful when a multiple supply voltage strategy is used in the design or when a larger single supply voltage is needed to drive the circuits. A multiple supply voltage strategy is required when level shifters are utilized in the circuit design in order to increase the robustness of the logic gate [94], [124], [125]. Dual-gate circuits are another option that requires an additional supply voltage, as discussed in the next chapter. At the current state of the art, three DC-DC converter implementations have been demonstrated in organic technologies.

The first implementation is a three-stage Dickson DC-DC converter, fabricated on a silicon substrate using single gate p-TFTs [126]. The maximum output voltage was 75 V with an output power of 0.5 mW, supplied by a VDD of 20 V. Subsequent implementations deliver a positive and negative maximum output voltage and are fabricated using a dual-gate p-TFT technology. There are only small design variations between the designs. At a supply voltage of 20 V, one DC-DC converter delivers +50 V and −40 V maximum output voltages with an output power of 0.17 µW [127]. The other implementation resulted in +48-V and −33-V maximum output voltage with an output power of 0.66 µW [128].

In display design, source drivers in the display periphery would require a digital-to-analog converter (DAC) fabricated in the same technology to deliver the signals containing the image content [129]. Xiong et al. demonstrated in 2010 the first 6-bit DAC realized in a complementary organic technology with a minimum feature length of 20 µm [62]. The supply voltage was as low as 3 V, yielding a maximum sampling rate of 100 S/s. This design has been realized with a C-2C DAC approach because the variation on capacitors is much less compared to the spread in transistor current. In subsequent years, two more 6-bit DACs were published, based on the current-steering principle. First, in 2011, Zaki et al. implemented this DAC in a p-type organic transistor technology with a minimum feature length of 4 µm [130]. Powered by a supply voltage of 3.3 V, it can convert at a maximum sampling rate of 100 kS/s. The layout of this chip severely reduces the parasitic wiring capacitances. Raiteri et al. have exceeded this maximum sampling rate to 10 MS/s at a VDD of 3 V [131]. This current-steering DAC has been realized in a-IGZO technology with a device channel length of 5 µm.

Analog-to-digital converters (ADC) are key circuits that function as the interface between the analog world (such as sensor nodes) and digital circuits (such as RFID transponder chips or microprocessors). Only two implementations have been demonstrated at the current state of the art, both published in 2010 [61], [64], [123]. Xiong et al. realized a C-2C successive-approximation ADC based on the C-2C DAC discussed previously [123]. The digital logic in this work is executed by an external FPGA. The analog part comprises 27 organic p-TFTs, 26 organic n-TFTs, and 19 metal-to-metal capacitors. Marien et al. published a fully integrated sigma-delta ADC, fabricated in a dual-gate organic p-TFT technology [64], [61]. This design employs 129 transistors.

2.6   Summary

In this chapter, we reviewed organic and metal-oxide thin-film transistor technologies and their applications to circuits. We first discussed different device configurations. The bottom-gate coplanar configuration is the main device layout for circuit realizations in this dissertation.

Organic and metal-oxide TFTs operate in accumulation and can be fully depleted because of the thin semiconducting layer. They can be biased in both linear and saturation regimes. These transistors are mostly unipolar, in other words mostly favor either electron or hole charge transport. This impacts their ability to be integrated in circuits, as we will discuss in coming chapters. Several methods have been found to tune the threshold voltage and thus enable more robust digital and analog circuits. Dual-gate transistors enable a promising route toward robust design. Each transistor can have its individual second gate used as a V_T-control gate or back gate. These dual-gate transistors can be used to implement stable unipolar, dual-V_T digital logic circuitry. More details can be found in the next chapters. This additional gate can also be connected to the front gate, resulting in an increased transconductance, and can be used in analog circuits to improve the gain. One drawback for most organic and metal-oxide TFT technologies is that they are not self-aligned and hence suffer from relatively large parasitic capacitances C_{gs} and C_{gd}. These will limit the performance of state-of-the-art circuit realizations.

Subsequently, we gave an overview of all six technologies used in this work for circuit realizations in Chapters 5 and 6. The first two technologies have been developed by Polymer Vision and are based on a 350-nm organic gate dielectric. The first is a single-gate, organic p-type technology; the second technology has the availability of a back gate. Next, an organic technology comprising a 100-nm high-k Al_2O_3 gate dielectric was explored in order to investigate channel length downscaling and overlap capacitance reduction. An increased circuit performance can be achieved by use of a-IGZO as semiconductor. Finally, two hybrid organic/metal-oxide, complementary technologies have also been discussed, with great potential for very robust and good performing circuit realizations.

The last section in this chapter gave an outline of trends in circuit integration for three different circuit topics, namely, display periphery circuits, digital circuits, and analog circuits. A main trend in display periphery is to use the same backplane technology (in many cases unipolar) for circuit integration. For digital and analog circuits, more complex chip integration is realized using more robust logic gates: from unipolar to dual-gate to complementary logic. Four trends are observed aiming for faster digital circuits in these technologies. The first is to implement the logic gates in a diode-load configuration instead of zero-V_{GS}-load, realized in a dual-gate technology. Other trends are the integration of higher-performance semiconducting materials such as metal-oxides, the downscaling of channel length and parasitic capacitance, and the fabrication of complementary circuits. Complementary circuit technologies result also in a decrease in supply voltage. The same holds for dual-gate circuits. Dielectric thickness scaling to a few nanometers only can yield circuits operating at supply voltages down to 2 V. For analog circuits, three relatively complex circuit blocks have been explored, namely, DC-DC converters enabling an additional supply, DAC converters for source drivers in displays, and ADC converters as an interface between the digital and analog worlds. Trends for analog circuits are similar to those for digital circuits.

3   Basic Gates

Digital circuits process digital information, represented in most cases by two voltage levels, one near the supply voltage (logic "1") and one near the ground voltage (logic "0"). Digital building blocks can perform simple Boolean functions such as an inverter or a NAND but can also provide more complex functions, such as a full adder, or include volatile memory blocks such as flipflops. In this chapter, we will focus our discussion on an inverter. The design of an inverter can easily be extended to more complex gates such as NANDs and NORs. In terms of inverters and NANDs any digital function can in principle be built. This results in a suboptimal solution. Optimal designs require the availability of more diverse and complex logic gates. Many organic and metal-oxide thin-film technologies on foil offer only one semiconductor type. Figure 3.1 depicts the scheme of an inverter and a 2-input NAND in complementary technology.

In the first section of this chapter, we will introduce important figures-of-merit for basic gates. The following section will describe the characteristics of inverters based on different topologies, depending on the availability of thin-film transistor technologies. We will elaborate in detail on zero-V_{GS}-load logic, diode-load logic, and complementary logic. The approach to extract these characteristics is similar as the one described in [132]. This chapter will end with a summary of the properties of these logic families and suggestions for improvements. For more explanations and details on technologies discussed in following sections, we refer to Chapter 2.

3.1   Figures-of-Merit

Figure 3.2 depicts some figures-of-merit of an inverter stage, more precisely the trip voltage (V_M), gain, and noise margin. The noise margin, described in the late 1960s [133], measures the immunity of an inverter against inevitable noise picked up by the circuit and added to the input voltage. As noise at the input gate of an inverter or any other logic gate and a V_T variation of the corresponding transistor result in the same current variation through the device, the noise margin is a very valuable parameter to measure the sensitivity of the circuit to threshold voltage variation.

There are many definitions to quantify the noise margin (also called static noise margin [SNM]), but a convenient one is the *maximum equal criterion,* or MEC

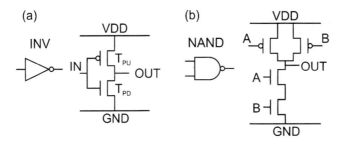

Figure 3.1   Implementation of (a) an inverter and (b) a 2-input NAND in complementary technology and their symbol.

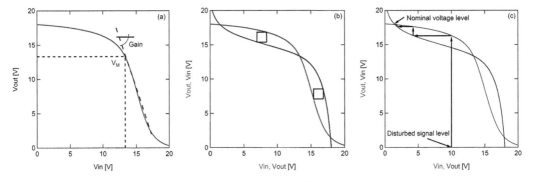

Figure 3.2   (a) Voltage transfer curve of the inverter circuit, with definition of trip voltage and gain; (b) determination of the noise margin of the inverter according to the MEC principle between the original and mirrored VTC; (c) demonstration of a regenerative gate.

[134]–[136]. According to this criterion, the noise margin is extracted by mirroring input and output voltages in the voltage transfer curve (VTC) and determining the size of the maximum square that fits between both curves, schematically represented in Figure 3.2 (b).

The trip voltage of the inverter is the data point on the transfer curve at which the output voltage equals the input voltage, as depicted in Figure 3.2 (a). In an ideal inverter from a static point of view, the trip point is $VDD/2$. Another important parameter is the gain. A minimum condition for an inverter yielding (high) noise margins is that it would have gain at V_M, that is, a region with slope larger than unity in the VTC (see Figure 3.2 (a)).

Moreover, for an integrated circuit to become possible, the output of an inverter has to be able to serve as input for a subsequent inverter stage. If the signal in the circuit is disturbed by noise and therefore differs from the nominal voltage level, a regenerative gate (see Figure 3.2 (c)) ensures that this signal gradually increases to the nominal voltage level after a number of logical stages [132]. The ratio between the SNM and the voltage difference between the logic levels gives a good measure of how many inverters are needed to restore the nominal logic levels.

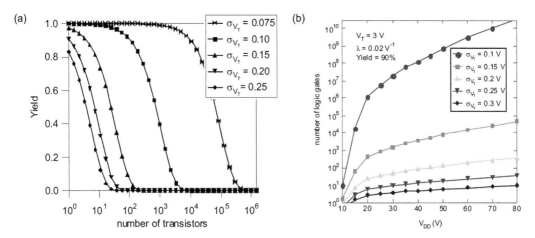

Figure 3.3   (a) Worst-case calculation of the circuit yield as a function of the number of zero-V_{GS}-load unipolar inverters, calculated for $VDD = 15$ V and an average V_T of 3 V; (b) the average noise margin increases for increasing VDD, such that the number of logic gates that can be integrated also increases (other parameters kept the same as for (a)).

The noise margin is a useful figure-of-merit to determine the yield of complex logic circuits integrating a large number of logic gates in the presence of parameter spread (e.g., on V_T and/or mobility). As an example, we show in Figure 3.3 the worst-case calculated yield as a function of the number of concatenated unipolar zero-V_{GS}-load inverter stages as shown in Figure 3.1 (a), with the variability in the threshold voltage as the parameter [137]. Yield is calculated as the joint probability that all inverter stages will have a noise margin larger than 0. The right panel shows that the number of stages that can be integrated for a certain yield, here 90 percent, increases with VDD thanks to the increasing average noise margin.

3.2   Logic Families

The inverter design in current silicon MOSFET technologies is based on a CMOS push-pull inverter, whereby the pull-up stage is a p-type transistor and an n-type transistor is used for the pull-down action (Figure 3.4 (a)). This design requires an approximate current matching between n- and p-type transistors and therefore a technology that is more complex compared to a unipolar technology (a process in which only one type of transistor is available). Current organic and metal-oxide technologies offer mostly unipolar transistors (best characteristics for p-types in organic and n-types in metal-oxide technologies), resulting in unipolar building blocks and circuits. The design of, for instance, a unipolar p-type inverter is shown in Figure 3.4 (b), whereby the load can be implemented as a resistor, for instance, using a conductive material such as PEDOT:PSS [138]. However, in order to avoid the integration of such extra material in the TFT process and because of the area

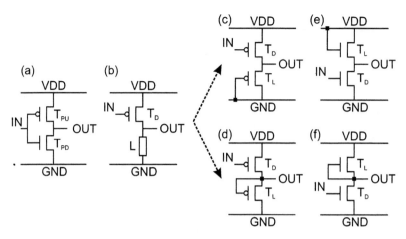

Figure 3.4   Different implementations for an inverter in (a) complementary logic and (b) unipolar p-type logic with a load L. This load is replaced by a p-type transistor that is connected as a (c) diode load or (d) zero-V_{GS} load. The n-type counterparts are depicted for (e) diode-load and (f) zero-V_{GS}-load inverters.

taken by the resistor, the load is most often implemented as a load transistor T_L (see Figure 3.4 (c) and (d)). This technique is well known for silicon p-channel MOS logic and later n-channel MOS logic in the 1970s. In silicon technology, one could use the doping of the semiconductor to define a different threshold voltage for the drive and load transistors ("dual-V_T" technology), while in organic TFT technology no reproducible method is known yet to control the threshold voltage of two adjacent transistors independently. Therefore, organic thin-film transistor logic is mostly unipolar p-type and single-V_T, while metal-oxide thin-film transistor logic is based on a unipolar, single-V_T, n-type technology. The unipolar n-type implementations for diode-load and zero-V_{GS}-load inverters are represented in Figure 3.4 (e) and Figure 3.4 (f), respectively.

3.3   Unipolar Logic

This section focuses on p-type only organic electronics. The methodologies and approaches used here can be extrapolated to other technologies, such as metal-oxide electronics (unipolar n-type).

Figure 3.4 (c–f) depict the two main possibilities for realizing unipolar logic inverters (and more complex gates), zero-V_{GS}-load logic or diode-load logic. Both inverters comprise a drive transistor, which is the pull-up transistor in the case of p-type logic, and a load transistor, here the pull-down transistor. Technologically this category of logic is the most convenient implementation, as a result of the presence of only one semiconductor and its patterning steps. Therefore, it is the most widespread technology to realize circuits in organic and metal-oxide thin-film technologies.

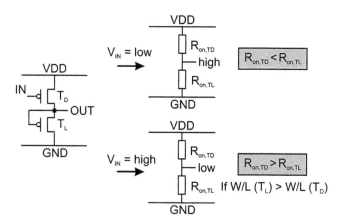

Figure 3.5   Operation principle of a zero-V_{GS}-load inverter, explained for a p-type technology.

On the other hand, the noise margin of unipolar inverters is limited by the poor gain of unipolar logic aggravated by the overall poor control and reproducibility of the threshold voltage of organic transistors [137]. This results in difficulties in placing the trip voltage at the middle rail ($VDD/2$) in a unipolar, single-V_T technology. The noise margin improves with increasing VDD; that explains why many organic circuits with a significant amount of transistors require substantial VDD to operate properly (see Figure 3.3 (b)). It also improves when the gate dielectric can be made very thin, culminating in gate dielectrics formed as a self-assembled monolayer on the gate electrode [139].

3.3.1   Single V_T, Depletion-Load, or Zero-V_{GS}-Load Logic

In zero-V_{GS}-load configuration, the gate and source nodes of the load-transistor are shorted (see Figure 3.5). Therefore, the load transistor results in a current source at a constant V_{GS}-voltage of 0 V. As a consequence, this transistor needs to be a depletion transistor, for example, a positive threshold voltage for the p-type transistor. The operation of the inverter is demonstrated in Figure 3.5. For low V_{IN}, the drive transistor is strongly on (large negative V_{GS}-voltage); in other words, the channel resistance of this transistor is low. The load transistor with positive V_T is not switched off, but leaking, whereby the channel resistance of the load transistor ($V_{GS} = 0$ V) is much larger than the equivalent resistance of the fully switched on drive transistor. Therefore, the resistive divider between the load and drive transistor sets V_{OUT} close to VDD. For more details on equivalent resistance definitions for transistors, we refer to Rabaey et al. [132].

In the other case, when V_{IN} is high, the drive transistor has a gate-source voltage close to 0 V, similar to the load transistor. Because of this and since the threshold voltages are similar, the size (W/L) of the load transistor should be chosen significantly (e.g., 5 to 10 times) larger than that of the drive transistor. Now, the resistive divider between the load and drive transistor pulls V_{OUT} close to GND. For these

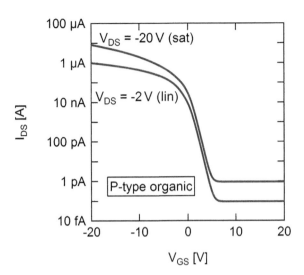

Figure 3.6 I_{DS}-V_{GS} curve (linear and saturation) of a p-type organic thin-film transistor.

inverters, non-idealities are present, being the swing of V_{OUT} that is smaller than 0 to VDD, caused by the leakage current of both the drive and load transistors, which are always switched on (depletion devices). Therefore, there is always a leakage current present in these inverters.

Equation (3.1) describes the behavior of the load and drive transistor when operating in a saturation regime. In a zero-V_{GS}-load inverter, the drain current of the load transistor quadratically depends on the threshold voltage (see Eq. (3.1)). This implies that any little variation in the threshold voltage will have an impact on the static inverter parameters, such as gain, switching threshold, and noise margin, but also on the dynamic inverter parameters.

$$\begin{cases} I_{SD,load} = \frac{1}{2}\mu C_{ox}\left(\frac{W}{L}\right)_{load}\left(V_{T,load}\right)^2 \\ I_{SD,drive} = \frac{1}{2}\mu C_{ox}\left(\frac{W}{L}\right)_{drive}\left(V_{SG,drive} + V_{T,drive}\right)^2 \end{cases} \tag{3.1}$$

3.3.1.1   VTC of the Zero-V_{GS}-Load Inverter

The VTC, or the voltage transfer curve, of the zero-V_{GS}-load inverter can be extracted from the output curves of the drive and load transistor. For a zero-V_{GS}-load inverter, the main requirement for the load transistor is that it should be a depletion device, which is common for most organic thin-film transistor technologies. Therefore, both transistors in this single-V_T, zero-V_{GS}-load inverter are depletion transistors. An example of such a depletion-load transfer curve is plotted in Figure 3.6 for both the linear and the saturation regime. These curves are simulation characteristics that are used to derive all parameters in this section.

Prior to the VTC extraction of the zero-V_{GS}-load inverter, we will transform the output curves of the load and drive transistors to a common coordinate set. This

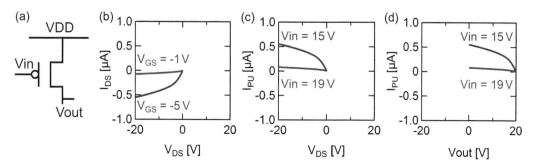

Figure 3.7 Transforming the output curve of the drive transistor into load curves for the inverter.

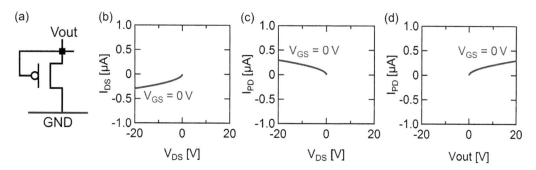

Figure 3.8 Transforming the output curve of the load transistor into a load curve for the inverter.

is depicted schematically in Figures 3.7 and 3.8 for the drive and load transistor, respectively. The scheme of the drive transistor is shown in Figure 3.7 (a). The source is always connected to VDD; its gate is the input voltage, and the drain results in the output voltage. Figure 3.7 (b) shows the output curve of the drive transistor for only two gate-source voltages for clarity reasons. We can list now the following equations:

$$V_{IN} = VDD + V_{GS} \tag{3.2}$$

$$I_{DS} = - I_{PU} \tag{3.3}$$

When we perform these transformations for VDD of 20 V, Figure 3.7 (c) can be plotted. The next operation is to plot the drain-source or pull-up current as a function of the output voltage. For this, we can use following equation:

$$V_{OUT} = V_{DS} + VDD \tag{3.4}$$

These transformations result in Figure 3.7 (d), where the pull-up current is plotted versus the output voltage for different input voltages.

A similar route can be followed to obtain the same curve for the load transistor. The source of the load transistor is the output node, which is also connected to

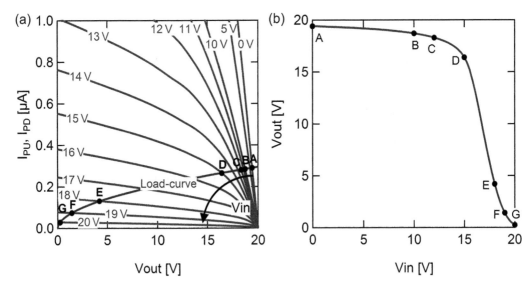

Figure 3.9   Load curves of the load and drive transistor and subsequent extraction of the inverter transfer curve.

its gate. The drain of this transistor is connected to the circuit's ground. The load line for the load transistor will remain constant, since the V_{GS}-voltage does not change. Here, the equation needed to transform the output curve Figure 3.8 (b) into Figure 3.8 (c) is

$$I_{DS} = -I_{PD} \tag{3.5}$$

Next, the pull-down current can be plotted as a function of the output voltage using

$$V_{OUT} = -V_{DS} \tag{3.6}$$

Figure 3.9 plots the load curves of both the load and drive transistors in the same coordinate set. The *W/L* ratio between the load and drive transistor chosen in this example is a 10:1 ratio. The input voltages are varied between 0 V and 20 V. Each intersection of the load curves of the load and drive transistor leads to a DC operation point of the inverter's VTC, because the pull-down and pull-up currents are matched at each DC operation point. For example, when the input voltage equals 0 V, the subsequent intersection point is depicted with "A." The output voltage corresponding to "A" can be read out and plotted in the VTC (Figure 3.9 (b)).

Varying the *W/L* transistor ratio will have an influence on the VTC, shown in Figure 3.10. An increase of the *W/L* size of the load transistor leads to larger load currents, which as a consequence shift the VTC to the left. This makes the *W/L* ratio a design parameter used to shift the VTC curve to a desired situation where, for example, the gain could be maximized at the trip voltage.

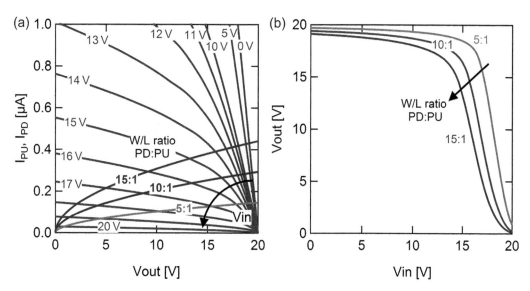

Figure 3.10   Influence of W/L transistor ratio between load and drive transistor on (a) the load curves and (b) VTC of the zero-V_{GS}-load inverter.

3.3.1.2   Static Parameters of the Zero-V_{GS}-Load Inverter

V_M

The trip point or switching threshold is defined at the VTC where V_{IN} equals V_{OUT}. We will for these calculations assume that VDD is large enough that both the drive and load transistors are operating in saturation regime (as shown in Figure 3.9). Moreover, channel length modulation effects are ignored. To derive the equation for V_M, we start observing that the pull-up and pull-down currents are equal. By substituting the equation for saturation regime, we obtain Eq. (3.8).

$$I_{PU} = I_{PD} \tag{3.7}$$

$$\frac{1}{2}\mu C_{ox}\left(\frac{W}{L}\right)_{PU}\left(V_{SG_{PU}}+V_{T,PU}\right)^2 = \frac{1}{2}\mu C_{ox}\left(\frac{W}{L}\right)_{PD}\left(V_{SG_{PD}}+V_{T,PD}\right)^2 \tag{3.8}$$

In this equation, we can assume that C_{ox} is equal for both the load and drive transistors. Apart from variability on μ and threshold voltages (see next chapter), we assume that the average μ and threshold voltages are equal for both transistors in single-V_T technology. Next, V_M is introduced in the equation:

$$\left(\frac{W}{L}\right)_{PU}\left(VDD-V_M+V_{T,PU}\right)^2 = \left(\frac{W}{L}\right)_{PD}\left(V_{T,PD}\right)^2 \tag{3.9}$$

Solving (3.9) leads to the equation for V_M for this inverter type:

$$V_M = VDD - \sqrt{r}\left(V_{T,PD}\right) + V_{T,PU} \qquad whereby\ r = \frac{\left(W/L\right)_{PD}}{\left(W/L\right)_{PU}} \tag{3.10}$$

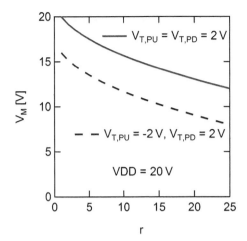

Figure 3.11   Influence of the W/L transistor ratio between the load and drive: transistor on switching threshold for (solid line) depletion devices with equal threshold voltage and for (dotted line) a different threshold voltage of both transistors.

We can conclude from the latter equation that the ratio r between the drive and load transistor is the main parameter to influence the switching threshold in this inverter type when the threshold voltages are equal. However, another parameter to tune could be a different threshold voltage for the load and drive transistor, if technology allows. Of course it should be noted that, for this inverter type, the load transistor should remain a depletion device. Figure 3.11 demonstrates the influence of the ratio on the switching threshold for single- and dual-V_T devices. For the single-V_T case, V_M is asymmetric and close to VDD but can be decreased by increasing the W/L ratio. These findings correspond to Figure 3.10. Hence, adding a second threshold voltage could also lead to a decrease of V_M but requires a unipolar technology with two threshold voltages. This additional threshold voltage would also allow the use of a ratio of 1. The next section discusses a valuable option to have two different V_T's by the addition of a back gate (see Section 3.3.2).

Gain

In digital design, the gain at the switching threshold is an important parameter to investigate the robustness of the inverter against variations. The larger the gain, the more abrupt change between high and low voltage is obtained. This yields subsequently larger noise margins. In order to retrieve a formula for the gain at V_M, small signal analysis of the inverter can be performed around V_M. At V_M, we assume that both transistors are operating in saturation regime (like Figure 3.9). Small signal analysis is performed at a specific DC operation point where a small signal is superposed on top. In this analysis, we investigate the behavior of the circuits caused by this small signal variation in the DC operation point.

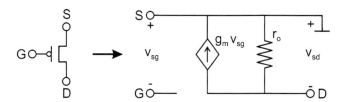

Figure 3.12   Small signal equivalent circuit of a p-type transistor.

The following notation is used:

- V_{IN} stands for the DC signal
- v_{in} stands for the small AC signal
- v_{IN} is the total signal

The total signal equals a DC signal on which a small AC signal is added.

$$v_{IN} = V_{IN} + v_{in} \tag{3.11}$$

The small signal equivalent of a p-type transistor is drawn in Figure 3.12.

The following equation describes the small-signal drain current of an organic thin-film transistor:

$$i_d = g_m v_{sg} + \frac{1}{r_o} v_{sd} \tag{3.12}$$

whereby

$$g_m = \left. \frac{\partial i_d}{\partial v_{gs}} \right|_{V_{GS} = V_{GS0}} \tag{3.13}$$

and

$$g_o = \frac{1}{r_o} = \left. \frac{\partial i_d}{\partial v_{ds}} \right|_{V_{DS} = V_{DS0}} \tag{3.14}$$

The transconductance g_m describes the variation in drain current caused by a small variation on the input voltage v_{gs} at the DC operation point for constant drain-source voltage.

The output conductance g_o describes the variation in drain current caused by a small variation on the drain-source voltage v_{ds} at the DC operation point for constant gate-source voltage. The output conductance is inversely proportional to the output resistance of the transistor. For a transistor in saturation mode, the output conductance is zero for ideal curves. The resulting source-drain current for practical transistors is influenced by the channel length modulation factor λ times the applied source-drain voltage (Eq. (3.15)), as explained in Section 2.2.1 of the previous chapter. λ is inversely proportional to the channel length L: when the transistor's channel length gets larger, the output conductance decreases.

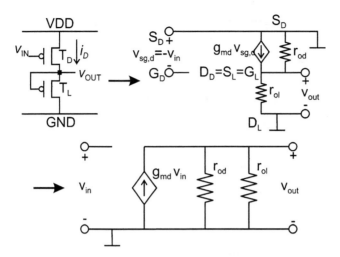

Figure 3.13   Small signal circuit equivalent of the zero-V_{GS}-load inverter.

Overall, the output conductance will be smaller compared to the transconductance, since the gate-source voltage dependency on the source-drain current is larger than the source-drain voltage dependency.

$$I_{SD} = \frac{1}{2}\mu C_{ox}\frac{W}{L}\left(V_{SG}+V_T\right)^2\left(1+\lambda V_{SD}\right) \tag{3.15}$$

Figure 3.13 shows the method to obtain the small signal circuit equivalent in the case of the zero-V_{GS}-load inverter. If we sum the currents at the output node of Figure 3.13 (bottom), we obtain the following equation:

$$g_{md}v_{in} + g_{od}v_{out} + g_{ol}v_{out} = 0 \tag{3.16}$$

Equation (3.16) can then be solved to obtain the (small signal) gain at V_M:

$$A_{V_M} = \frac{v_{out}}{v_{in}} = -\frac{g_{md}}{g_{od}+g_{ol}} \tag{3.17}$$

Because the transconductance yields larger values compared to the output conductance, the absolute value of the gain at V_M is sufficiently larger than unity.

3.3.1.3   Dynamic Behavior of the Zero-V_{GS}-Load Inverter

The transient analysis of an inverter can be estimated from the analysis of a first-order RC network [132]. The resistance values can be related to the on-resistance of the transistors, while the capacitances are the load capacitors. We assume in this analysis that the inverter has its trip point around half the supply voltage. The first-order approximation for pull-down delay and pull-up delay is [132]

$$\begin{cases} t_{pHL} = \ln(0.5)\,R_{eql}C_L = 0.69R_{eql}C_L \\ t_{pLH} = \ln(0.5)\,R_{eqd}C_L = 0.69R_{eqd}C_L \end{cases} \tag{3.18}$$

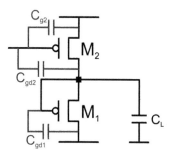

Figure 3.14 Capacitors of a zero-V_{GS}-load inverter.

The average delay time of the inverter is defined by the following equation:

$$t_p = 0.69 C_L \left(\frac{R_{eql} + R_{eqd}}{2} \right) \tag{3.19}$$

In zero-V_{GS}-load inverters, the pull-down time is much longer than the time to pull up. This is due to the large value for on-resistance of the load transistor, which has its gate connected to its source. One could now argue that increasing the size of the load transistor would lead to faster inverters. In one sense, this is indeed correct, since the equivalent resistance of the load transistor would increase; however, we also have to take into account that the load capacitance of the zero-V_{GS}-load transistor also increases.

Study of the Zero-V$_{GS}$-Load Capacitors
Figure 3.14 shows schematically all parasitic capacitors of a zero-V_{GS}-load inverter. The definitions of these capacitors are also listed in Table 3.1. Please note the α_1 and α_2 parameters for the channel capacitances in this table. For the calculations further in this chapter, we have estimated the gate-to-source channel capacitance parameter α_1 to 2/3 and removed the gate-to-drain channel capacitance contribution for a transistor operating in saturation regime [140], [141] $(\alpha_2 = 0)$.

In most current metal-oxide and organic thin-film technologies, the source-drain electrodes have to be laid out fully overlapping with the gate electrode. This is an intrinsic limitation due to the presence of a non-self-aligned gate technology. These limitations play an important role in the overall operational speed of these circuits, but are also limiting the RC delay in large area displays based on organic or metal-oxide transistor backplanes. It is clear that there is a major focus in technology research nowadays targeting self-aligned thin-film transistors to deal with these issues [65].

The value for gate-drain and gate-source overlap capacitances can be calculated by Eq. (3.20):

$$\begin{cases} C_{gdx} = C_{ox} \left(A_{gdoverlap} \right)_x \\ C_{gsx} = C_{ox} \left(A_{gsoverlap} \right)_x \end{cases} \tag{3.20}$$

Table 3.1. Definitions of All Parasitic Capacitors of a Zero-V_{GS}-Load Inverter

Capacitor	Definition	Value
C_{g2}	Gate capacitance of transistor M_2	$C_{g2} = C_{gso2} + C_{gcs2}$
C_{gso2}	Gate-source overlap capacitance of transistor M_2	$C_{gso2} = C_{gso2}$
C_{gcs2}	Gate-to-source channel capacitance of transistor M_2	$C_{gsc2} = \alpha_1 C_{ox} (WL)_2$
C_{gd2}	Gate-drain capacitance of transistor M_2	$C_{gd2} = C_{gdo2} + C_{gcd2}$
C_{gdo2}	Gate-drain overlap capacitance of transistor M_2	$C_{gdo2} = C_{gdo2}$
C_{gcd2}	Gate-to-drain channel capacitance of transistor M_2	$C_{gcd2} = \alpha_2 C_{ox} (WL)_2$
C_{gd1}	Gate-drain capacitance of transistor M_1	$C_{gd1} = C_{gdo1} + C_{gcd1}$
C_{gdo1}	Gate-drain overlap capacitance of transistor M_1	$C_{gdo1} = C_{gdo1}$
C_{gcd1}	Gate-to-drain channel capacitance of transistor M_1	$C_{gcd1} = \alpha_2 C_{ox} (WL)_1$

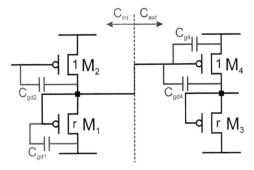

Figure 3.15   Intrinsic and extrinsic capacitors of a zero-V_{GS}-load inverter.

The load capacitance of an inverter can be attributed to its intrinsic output capacitor and its extrinsic capacitor. The latter capacitor is the combination of the wiring capacitor and the fanout of the inverter. For further analysis in this section, we will ignore the wiring capacitor. Figure 3.15 outlines the intrinsic and extrinsic capacitors of a zero-V_{GS}-load inverter. It should be noted that the ratio r should be larger than unity (see previous section, e.g., V_M-analysis). The intrinsic capacitor of the inverter is defined by the gate-drain capacitance of the drive transistor and the gate-channel and gate-drain capacitances of the load transistor, which are approximately r times larger than the drive transistor. On the other hand, the extrinsic capacitor is the sum of all parasitic capacitors related to the drive transistor. This is also written down in Eq. (3.21).

$$\begin{cases} C_{int} = 2C_{gd2} + C_{gd1} \\ C_{ext} = C_{g4} + C_{gd4} \end{cases} \tag{3.21}$$

C_{gd2} should be included with a multiplication factor taking the overshoot at the output node of the first inverter into account, also known as the digital Miller effect. An example of this overshoot is depicted in Figure 3.16 (a) for the situation when the

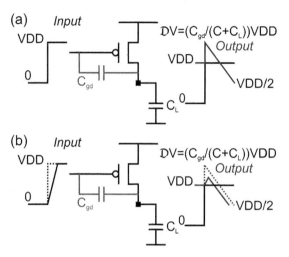

Figure 3.16 Overshoot on the output node of an inverter consequence of C_{gd} when the input signal varies between *GND* and *VDD* for (a) zero rise time and (b) non-zero rise time.

applied input signal becomes *VDD*. The absolute value of the overshoot can be calculated by the capacitive division between C_{gd} and all the other capacitors C_L, as shown in Figure 3.16. The time to reach *VDD*/2 can be estimated according to Eq. (3.22). This equation demonstrates that, theoretically, C_{gd} should be added a maximum of three times for zero rise and fall times. Nevertheless, we have opted for a multiplication factor of 2 since zero rise and fall times will not occur. The effect of a larger rise time is shown in Figure 3.16 (b), resulting in smaller overshoots. From Eq. (3.21), we can now estimate that C_{int} is larger than C_{ext} when the ratio r becomes larger than or equal to 1.

$$t_d = \frac{C_{total}\left(\Delta V + \dfrac{VDD}{2}\right)}{I} = \frac{\left(C_{gd}+C_L\right)\left(\dfrac{C_{gd}}{C_{gd}+C_L}VDD+\dfrac{VDD}{2}\right)}{I}$$

$$= \frac{\left(3C_{gd}+C_L\right)VDD}{2I} \tag{3.22}$$

We will now substitute C_L with the intrinsic and extrinsic capacitors. This leads to Eq. (3.23):

$$t_p = 0.69 R_{eq}C = 0.69 R_{eq}\left(C_{int}+C_{ext}\right) = 0.69 R_{eq}C_{int}\left(1+\frac{C_{ext}}{C_{int}}\right) \tag{3.23}$$

We define the intrinsic delay, which is the delay loaded only with its intrinsic capacitance.

$$t_{d0} = 0.69 R_{eq}C_{int} \tag{3.24}$$

The next step is to determine the effect of inverter sizing on the delay. We take the minimum sized inverter as reference gate. When introducing S as size of the inverter, we can write that the R_{ref} will be decreased by this factor S, because the on-resistance of both load and drive transistors will decrease with the sizing factor. The intrinsic capacitance will be proportional to the sizing factor S, since they also depend on the geometrical size of both transistors. Therefore, we can write

$$t_p = 0.69 \frac{R_{ref}}{S} C_{iref} S \left(1 + \frac{C_{ext}}{C_{iref} S} \right) = t_{d0} \left(1 + \frac{C_{ext}}{C_{iref} S} \right) \tag{3.25}$$

It can be seen that the intrinsic delay is not dependent on scaling. This is correct if indeed both the minimum resistance and the capacitance scale with S. Furthermore, increasing S to very high values could consequently lead to stage delays close to the intrinsic delay. In this equation, no external load capacitance is taken into account. This increase of S to very high values would also result in very large inverter footprints. Moreover, increasing S leads to a larger input capacitance of this inverter, which implies that the gate that drives this inverter will suffer from a big load. The intrinsic capacitor is larger than the extrinsic capacitor for a 1:10 transistor ratio as a consequence of the contribution of C_{gd1}. A transistor ratio of 1:1 results in a smaller intrinsic capacitance compared to the extrinsic capacitance.

The single-V_T, zero-V_{GS}-load inverter suffers intrinsically from the self-loading effect, or, in other words, the intrinsic capacitance is mostly dominant. Sizing of these inverters will as a consequence not help to yield shorter delays with certain loads, because scaling will also result in a larger intrinsic load capacitance, which will be added to the total capacitance needed to charge/discharge.

Hence, it is wise to have the intrinsic capacitance value as low as possible. When the transistor ratio is chosen closer to 1, the intrinsic capacitance becomes smaller than the extrinsic capacitance. Such transistor ratios can only be realized – taking into account the DC analysis – when two threshold voltages are available. A valid option for this is discussed in Section 3.3.2 (dual-gate zero-V_{GS}-load logic).

As a consequence, it is more relevant to design for a certain delay in a chain of logic gates (e.g., in the critical path). Targeting this, we have to sum up all delays of the logic gates in the chain.

$$t_p = t_{d0} \left(1 + \frac{C_{ext}}{C_{int}} \right) = t_{d0} \left(1 + \frac{f}{\gamma} \right) \tag{3.26}$$

The definition of proportionality factors γ and f can be found in the next equation.

$$\begin{cases} C_{int} = \gamma C_g \\ C_{ext} = f C_g \end{cases} \tag{3.27}$$

The input capacitor of the next inverter (or number of inverters/building blocks) multiplied by γ leads to the intrinsic capacitor. The extrinsic capacitor is the input

Table 3.2. Calculated and measured γ factor for zero-V_{GS}-load inverter with the following dimensions: W/L equals 140/5 and 1400/5, designed in the technology described in Chapter 2, Sections 2.4.1 and the channel capacitances are chosen for transistors operating in saturation mode

Ratio	A_{gd2} (μm^2)	A_{gs2} (μm^2)	A_{gsc2} (μm^2)	A_{gd1} (μm^2)	γ	$\gamma_{measured}$
1:1	500	1725	462	500	0.56	No data
1:10	500	1725	462	4285	1.97	2.02

capacitor of the next inverter (or number of inverters/building blocks) multiplied by the effective fanout f. Equation (3.26) demonstrates that the delay of the inverter is only a function of the ratio between its external load capacitance and the input capacitance.

Substituting the capacitors written in Eq. (3.21) leads to estimation of the proportional factor γ:

$$C_{int} = 2C_{gd2} + C_{gd1} = \gamma C_g = \gamma \left(C_{g2} + C_{gd2} \right) \tag{3.28}$$

yielding

$$\gamma = \frac{2C_{gd2} + C_{gd1}}{C_{gd2} + C_{g2}} = \frac{2C_{gd2} + C_{gd1}}{C_{gd2} + C_{gso2} + C_{gsc2}} \tag{3.29}$$

The proportionality factor (or technology factor) γ depends on the ratio between load and drive transistor and technology/layout metrics.

We can elaborate more practically on this for typical thin-film transistor layouts used in this work, as explained in the previous chapter, in Section 2.3. The base technology for this analysis is described in Sections 2.4.1 and 2.4.2. From this layout we calculated the transistor capacitances and subsequently the technology factor, listed in Table 3.2 for two different W/L transistor ratios. This γ factor is 1.97 for a 1:10 transistor ratio, while a 1:1 ratio results in a γ factor of 0.56, which is beneficial for delay optimization. Moreover, we have realized 19-stage ring oscillators whereby the driving stage is a single inverter; however, the load of each stage contains a number of inverters in parallel, as shown in Figure 3.17 (a). This circuit allows prediction of the technology factor γ and the intrinsic delay t_{d0} from Eq. (3.26). Figure 3.17 (b) plots the obtained measured delay as a function of the increased load for ratio 10. The black dotted line is the linear fit, which corresponds well with the measured data. The intrinsic delay can be extracted to be 3.43 μs (where the fanout equals zero) and γ is 2.02. The measured technology factor corresponds well to the estimated technology factor. Reasons for the small difference may be the Miller effect, for which we arbitrary took a factor of 2, and the channel capacitances of the load and drive transistor. These capacitances vary during each cycle, since the transistor's channel will switch from saturation mode to linear and eventually to subthreshold regime.

It is also clear from Table 3.2 and Eq. (3.29) that our asymmetrical layout (more source contacts than drain contacts) is advantageous for the final technology factor.

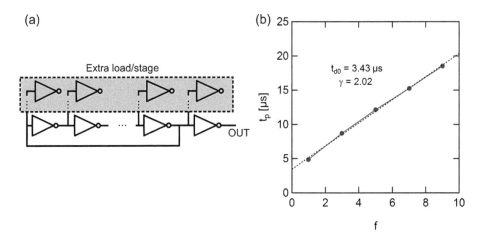

Figure 3.17   (a) Scheme of a 19-stage ring oscillator, including additional loads per stage, and (b) obtained measurements of a zero-V_{GS}-load inverter (r = 10) with different loads and its fitted curve (dotted line).

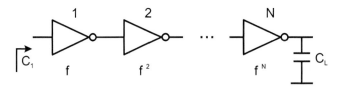

Figure 3.18   Chain of N inverters with a minimal input load C_I and an output load capacitor C_L.

This factor will be worse when the transistor is laid out symmetrically or has more drain contacts with respect to source contacts, especially for the 1:10 ratio.

In a real integrated circuit, we could have a chain of N inverters that have to drive a certain load, C_L. We are now interested in the minimal delay that can be obtained in order to drive this load with N inverters. Moreover, it should result in a minimal number of inverter stages. This is schematically depicted in Figure 3.18.

The total delay of this chain is the sum of all individual delays. The optimum size of each inverter is the geometric mean of its two neighbor sizes [132]:

$$C_{g,j} = \sqrt{C_{g,j-1}C_{g,j+1}} \tag{3.30}$$

Therefore, we can state that each stage has the same effective fanout (f) and consequently the same delay.

$$f = \sqrt[N]{\frac{C_L}{C_{g,1}}} = \sqrt[N]{F} \tag{3.31}$$

F is the equivalent fanout when driven by an individual inverter. The fanout of each stage is depicted in Figure 3.18. The minimum delay that can be calculated for the total chain is given by

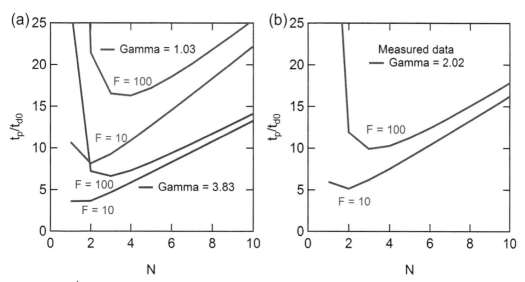

Figure 3.19 $\dfrac{t_p}{t_{d0}}$ plotted as a function of the number of inverter stages for different γ and overall effective fanouts for (a) calculated data and (b) measured data.

$$t_p = N t_{d0}\left(1 + \frac{\sqrt[N]{F}}{\gamma}\right) \tag{3.32}$$

In Eq. (3.32), there is a trade-off in the number of stages. If N becomes too large, the intrinsic delay times the number of stages becomes dominant. On the other hand, when N becomes too small, the overall effective fanout (F) of each stage becomes large. This is also shown in Figure 3.19 for different technology factors described in Table 3.2. There is a clear difference between the optimum number of stages for both different inverter sizes (1:1 vs. 1:10 ratio) for different values of overall effective fanouts. Additionally, Figure 3.19 (b) depicts the normalized delay for the measured technology factor for ratio 10. These figures show clearly that lower values of γ are beneficial for stage delay optimization. This can be explained by the fact that in this technology, C_{int} is substantially larger than C_{ext}, implying that the inverters suffer from the self-loading effect for large ratios. As a consequence, also from a dynamic point of view, a 1:1 transistor ratio for zero-V_{GS}-load inverters in these technologies is the preferred option.

The optimum value for effective fanout can be found by differentiating the equation for minimum delay (3.32) to the number of stages and setting the result to 0 [132]. We obtain the following recursive equation:

$$f_{opt} = e^{\left(1 + \frac{\gamma}{f_{opt}}\right)} \tag{3.33}$$

This equation is plotted in Figure 3.20. We can find all different f_{opt} values for the cases provided in Table 3.2. All are further listed in Table 3.3.

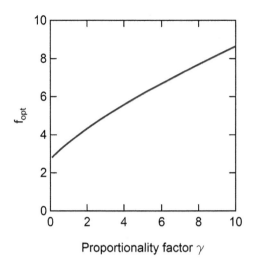

Figure 3.20   The optimum effective fanout plotted as a function of the proportional factor γ.

When we take the optimal f into account, we can calculate an optimal number of stages N. This number of stages will be an integer value in practice.

$$f_{opt}^{N_{opt}} = F \tag{3.34}$$

$$N_{opt} = \frac{\ln(F)}{\ln(f_{opt})} \tag{3.35}$$

Table 3.3 discusses the normalized optimum delay for two different optimum effective fanouts. Both effective fanouts are the values represented by two suggested transistor ratios (Table 3.2). For transistor ratio r equal to 1, there is a substantial difference between the options of unbuffered (or single stage), two stages, and an optimum inverter chain for increasing overall effective fanout. In contrast, for transistor ratio r equal to 10, the difference between unbuffered, two stages and the inverter chain becomes less compared to ratio 1. Only when the overall effective fanout becomes substantially large (driving the signal off the chip) is the use of optimum inverter chains highly recommended.

3.3.2   Dual V_T, Zero-V_{GS}-Load Logic by Dual-Gate TFTs

As we will conclude later in this chapter, the move toward a complementary technology is the most favorable route to obtain robust thin-film transistor circuitry and to increase the integration density. This requires, however, proper matching of p-type and n-type TFTs regarding device performance. This matching has turned out to be difficult for close integration in complex circuits for these thin-film technologies. Dual-V_T logic could be a valid alternative to robust circuitry for technologies on flexible substrates, as it was for dual-V_T unipolar Si-circuitry. This

Table 3.3. $\dfrac{t_p}{t_{do}}$ for various driver configurations

r	γ	f_{opt}	F	Unbuffered	Two stages	N_{opt}	Optimal chain
1	0.56	3.23	10	19.64	13.29	2 ~ round (1.96)	13.29
			100	180.36	37.71	4 ~ round (3.93)	26.59
			1000	1787.50	114.94	6 ~ round (5.89)	39.88
10	1.97	4.30	10	5.58	5.21	2 ~ round (1.58)	5.21
			100	51.27	12.15	3 ~ round (3.16)	10.07
			1000	508.12	34.10	5 ~ round (4.74)	15.10
10 (meas)	2.02	4.33	10	5.45	5.13	2 ~ round (1.57)	5.13
			100	50.00	11.90	3 ~ round (3.14)	9.89
			1000	495.54	33.31	5 ~ round (4.71)	14.85

option is validated by the realization of RFID transponder chips [115] and the first organic-transistor based microprocessor on foil [63].

In the previous section, we have investigated single-V_T, zero-V_{GS}-load logic for both its static and its dynamic properties. During this discussion, we have concluded several times that it would be beneficial to have two threshold voltages available for this logic family, because a ratio of 1:1 could then be used. One way to obtain multiple threshold voltages in a unipolar thin-film transistor technology has proven to be possible by the addition of a second gate in the technology. For the technology part, we refer to the previous chapter, where the principle of dual-gate transistors is explained. Figure 3.21 shows typical measured I_{DS}-V_{GS} characteristics of a dual-gate transistor [115]. By varying the back-gate bias between +30 V and −30 V, the transistors' threshold voltage can be controlled, leading from depletion-mode to more enhancement-mode curves, in agreement with earlier publications [89][90]. The extracted back-gate sensitivity factor in this technology is 0.2, as concluded in the previous chapter, Section 2.4.2.

3.3.2.1 VTC of a Dual-V_T Zero-V_{GS}-Load Inverter

We can derive the VTC of the dual-V_T zero-V_{GS}-load inverter from the load curves of the load and drive transistor. The output curve at V_{GS} equals −4 V, which has been chosen as the load curve for the load transistor. As stated before, we are using a favorable 1:1 transistor ratio between drive and load transistor. The corresponding VTC is shown as a full line in the right panel of Figure 3.22. The dashed line plots the VTC of the single-V_T, zero-V_{GS}-load inverter taken from Figure 3.9. This demonstrates again that similar VTC can be obtained for a dual-V_T zero-V_{GS}-load inverter.

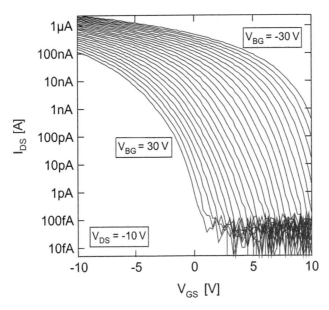

Figure 3.21 Typical measured transfer characteristic of a dual-gate transistor when using the back-gate as V_T-control gate. The channel width equals 140 μm, the channel length 5 μm.

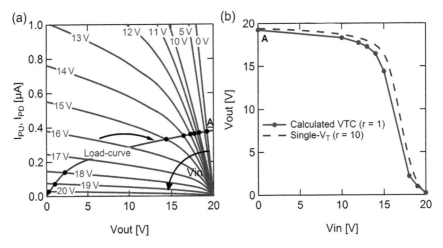

Figure 3.22 (a) Load curves of the load and drive transistor and (b) extraction of the VTC. The VTC of the single-V_T zero-V_{GS}-load inverter is added as a dashed line for comparison.

3.3.2.2 Dual-Gate Zero-V_{GS}-Load Inverter

Equation 3.10 and Figure 3.10 demonstrate that a second threshold voltage allows optimizing of the switching threshold of the inverter, resulting in the possibility of designing with a transistor ratio of 1. Practically, we have implemented this by replacing every single-gate transistor with a double-gate transistor, as shown in Figure 3.23 (a) and published in [115]. The V_T-control gate is denoted with the

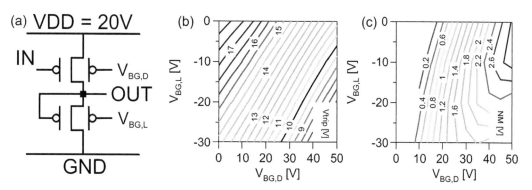

Figure 3.23 Scheme of a dual-gate zero-V_{GS}-load inverter (a). Contour plots of the trip voltage (b) and the noise margin (c) are determined from measured transfer curves of inverters when varying the back-gate voltage of the load and drive transistor and applying a supply voltage of 20 V.

subscript "BG." The ratio between the drive and load transistor for the zero-V_{GS}-load logic is designed with a 1:1 ratio. Figure 3.23 (b) and (c) depict contour plots of the noise margin and trip point when varying the V_T-control gate voltage of the drive and load transistor for both inverter topologies. As can be seen, noise margins can exceed 2.8 V for dual-gate zero-V_{GS}-load inverters at VDD of 20 V. Also, the trip point can be shifted toward $VDD/2$ (i.e., 10 V) by appropriate V_T control.

We have fabricated and measured 19-stage ring oscillators implementing these dual-gate zero-V_{GS}-load inverters with a 1:1 ratio [115]. Figure 3.24 shows the frequency obtained when varying the back gate of the drive and load transistor. A frequency range between 3 kHz and 24 kHz has been obtained. The influence of the threshold voltage of the drive transistor is much weaker than the influence of the threshold voltage of the load transistor, which delivers the small pull-down current to discharge the input capacitance of the next stage. The variation of the back gate is a viable route to improve the delay of the dual-gate zero-V_{GS}-load circuits.

3.3.2.3 Optimized Dual-Gate Zero-V_{GS}-Load Inverter

One disadvantage of the previous dual-gate configuration is that the threshold voltage of the load transistor changes dynamically with the output node, since this node serves also as the source contact for the back-gate transistor (see Figure 3.23). The threshold voltage moves in the opposite direction as required for high gains when the output node varies. In other words, the load transistor is made dynamically weaker when it should pull the output node down. Therefore, we have implemented an optimized architecture, as explained in Figure 3.25 [115]. This figure shows five measured VTCs for the dual-gate inverter when sweeping the back-gate voltage of the load transistor between *GND* and *VDD* and keeping the back-gate voltage of the drive transistor constant. The intersecting points at the VTC whereby the back gate of the load equals the output node are marked with triangles in Figure 3.25 (b). They form the VTC that would be obtained if

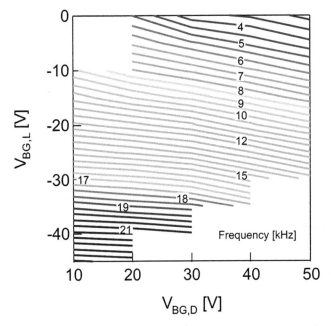

Figure 3.24   Contour plot of the frequency determined from measured curves of 19-stage ring oscillators using the dual-gate zero-V_{GS}-load architecture (Figure 3.23) when varying the back-gate voltage of the load and drive transistor and applying a supply voltage of 20 V.

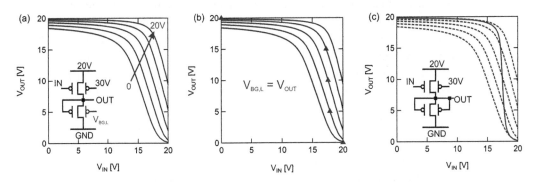

Figure 3.25   Dual-gate zero-V_{GS}-load inverter whereby the $V_{BG,L}$ is swept between 0 and 20 V and $V_{BG,D}$ is kept constant at 30 V: (a) shows the measured VTC curves; (b) the triangles are intersecting points at which the back gate of the load equals the output node; (c) the solid curve shows the measured VTC curve when the back gate of the load transistor is connected to the output node.

the output node were connected to the back gate of the load transistor. The solid curve in Figure 3.25 (c) shows the measured VTC of the optimized dual-gate inverter when the back gate of the load transistor is connected to the output node.

Figure 3.26 depicts the measured VTC of the dual-gate zero-V_{GS}-load inverter implemented in this optimized configuration [115]. The V_T-control gate of the

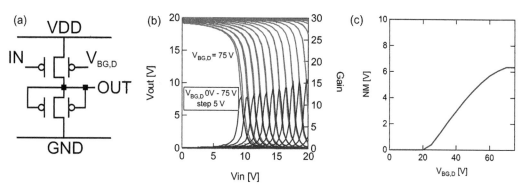

Figure 3.26 (a) Scheme of the optimized dual-gate zero-V_{GS}-load inverter, (b) the corresponding measured VTC while sweeping the back-gate voltage of the drive transistor and a *VDD* of 20 V, and (c) the extracted noise margin as a function of $V_{BG,D}$.

drive transistor can be used to move the trip point toward *VDD*/2. At this point, in zero-V_{GS}-load inverters, the gain exceeds 11 and the noise margin is larger than 6 V at *VDD* of 20 V. The decrease of the gain with respect to the increase of back-gate voltage of the drive transistor can be explained by the fact that g_{md} becomes smaller with increasing $V_{BG,D}$. Consequently, following Eq. (3.17), this lower g_{md} yields thus smaller values for the gain at V_M.

As shown in the previous figure, these excellent static properties enable integration into larger circuitry, such as RFID transponder chips [115] and the first organic 8-bit microprocessor on foil [63]. These realizations will be described in more detail in the following chapters.

For a dynamic perspective, we have already calculated the technology factor in the previous section; it is 1.04 as listed in Table 3.2. This optimum value for zero-V_{GS}-load logic is the consequence of a 1:1 ratio that can be provided by dual-gate technologies. As a conclusion, dedicated sizing of inverters with respect to different fanouts is more beneficial compared to single-V_T zero-V_{GS}-load logic with larger ratios. The impact is listed in Table 3.3 and plotted in Figure 3.19.

As a general conclusion for this section, we can state that this inverter family offers multiple advantages, regarding both static and dynamic properties, and is therefore a valid option for complex circuit realizations. Designs made in this technology are discussed further in the following chapters.

3.3.3 Single V_T, Enhancement-Load, or Diode-Load Logic

Figure 3.27 shows the configuration of a p-type, enhancement-load inverter. The gate and drain nodes of the pull-down transistor are connected, resulting in a diode-connected transistor. When the input voltage is high, the gate-source voltage of the drive transistor is small, yielding low currents, while the load transistor will be strongly on. In this situation, the output node will be pulled down. In the other case, when the input voltage is low, the gate-source voltages of both drive and load

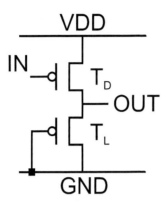

Figure 3.27   Schematic configuration of a p-type, diode-load connected inverter.

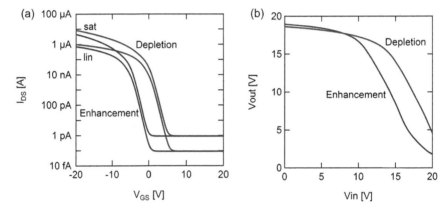

Figure 3.28   (a) Transfer curve (linear and saturation) of a p-type depletion and enhancement organic thin-film transistor and (b) VTC of a diode-load inverter having a 10:1 ratio (drive/load) for both depletion and enhancement devices.

transistors are similar. Therefore, in order to ensure that the output node is pulled up toward *VDD*, the dimensions of the drive transistor should be chosen significantly larger than those of the load transistor.

3.3.3.1   VTC of the Diode-Load Inverter

Figure 3.28 plots the VTC as extracted from simulations starting from depletion-mode transistors and enhancement-mode transistors. The enhancement-mode transistors result in correct operation of a diode-load inverter; however, also mildly depletion-mode transistors can yield functional logic inverters. The enhancement-mode transistors are the base for all derivations in this section of the chapter.

Similarly to the way we have derived the VTC of the zero-V_{GS}-load inverter, we can extract the VTC from the load curves of the load and drive transistor to

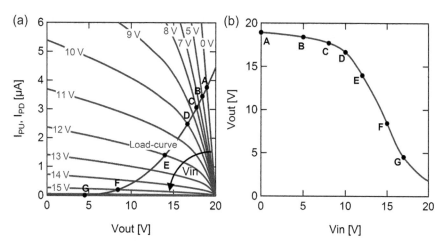

Figure 3.29   (a) Load curves of the load and drive transistor and subsequent extraction of the inverter transfer curve for a W/L ratio between the drive/load transistor of 10:1 and (b) the corresponding inverter VTC.

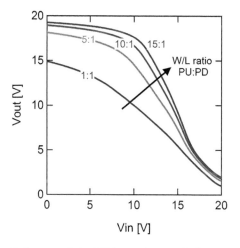

Figure 3.30   Influence of the W/L transistor ratio between the drive and load transistor on the VTC of the diode-load inverter.

find the DC operating point for each input voltage. Figure 3.29 shows the load curves and the corresponding VTC curve. Please note that the load transistor now is a diode-connected transistor. Moreover, we have chosen W/L dimensions for the drive transistor to be 10 times larger than the load transistor.

The influence of the W/L ratio between the drive and load transistor is shown in Figure 3.30. It is noteworthy that the drive TFT for the diode-load logic needs to be larger than the load-TFT, which results in opposite ratios compared to zero-V_{GS}-load inverters. In this figure, we can observe that increasing the W/L ratio leads to larger output swings and increased gain. A remarkable shift is already noted between 1:1 and 5:1 ratios.

3.3.3.2   Static Behavior of the Diode-Load Inverter

V_M

We will derive the equation for the trip voltage V_M using the same approach as for the zero-V_{GS}-load inverter, by describing the pull-up and pull-down currents.

$$I_{PU} = I_{PD} \tag{3.36}$$

$$\frac{1}{2}\mu C_{ox}\left(\frac{W}{L}\right)_{PU}\left(V_{SG_{PU}} + V_{T,PU}\right)^2 = \frac{1}{2}\mu C_{ox}\left(\frac{W}{L}\right)_{PD}\left(V_{SG_{PD}} + V_{T,PD}\right)^2 \tag{3.37}$$

In this equation, we can assume that C_{ox} is equal for both the load and drive transistors. Apart from variability on μ and threshold voltages (see next chapter), we assume that μ and threshold voltages are equal for both transistors in a single-V_T technology. The following equation will introduce V_M:

$$\left(\frac{W}{L}\right)_{PU}\left(VDD - V_M + V_{T,PU}\right)^2 = \left(\frac{W}{L}\right)_{PD}\left(V_M + V_{T,PD}\right)^2 \tag{3.38}$$

Solving (3.36) leads to the definition for V_M for a diode-load inverter:

$$V_M = \frac{VDD + V_{T,PU} - \sqrt{\frac{1}{r}}\left(V_{T,PD}\right)}{1 + \sqrt{\frac{1}{r}}} \quad \text{whereby } r = \frac{\left(W/L\right)_{PU}}{\left(W/L\right)_{PD}} \tag{3.39}$$

Please note that r is defined as opposite the zero-V_{GS}-load inverter. From Eq. (3.39), we conclude also here that the ratio is the main parameter to influence the trip point in this inverter type at a targeted supply voltage. The influence of the ratio on the switching threshold is shown in Figure 3.31. It increases strongly for small ratios, after which it starts to roll off. This is in agreement with Figure 3.30, where a big influence is also observed for small ratios. The threshold voltage of the load transistor has only limited influence on the trip point, which is in agreement with Eq. 3.39. This equation and Figure 3.31 (dashed lines) demonstrate that the threshold voltage of the drive transistor has a much stronger impact on the switching threshold.

Gain

The gain at V_M can be derived from the small signal analysis of the inverter. The representative small signal equivalent circuit of the diode-load inverter is shown in Figure 3.32. From this figure, we can extract the small signal gain for this inverter type in Eq. (3.40).

$$A_{V_M} = \frac{v_{out}}{v_{in}} = -\frac{g_{md}}{g_{od} + g_{ol} + g_{ml}} \tag{3.40}$$

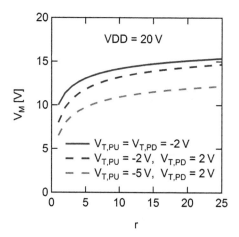

r

Figure 3.31 Influence of the *W/L* transistor ratio between the drive and load transistor on the switching threshold for (solid line) enhancement devices with equal threshold voltage and for (dotted lines) different threshold voltages of both transistors.

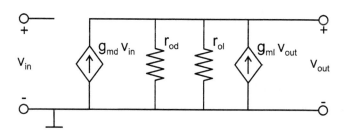

Figure 3.32 Small signal equivalent circuit of the diode-load inverter.

This formula shows that the gain for the equal transistor ratio is limited, but can be improved by increasing the transistor ratio between drive and load transistor. This is also in correspondence with Figure 3.30. One method to improve gain (and therefore noise margin) is by a dual-gate technology, discussed in Section 3.3.4.

3.3.3.3 Dynamic Behavior of the Diode-Load Inverter

Study of the Diode-Load Capacitances

Equation 3.41 describes the intrinsic and extrinsic capacitors of the diode-load inverter, illustrated in Figure 3.33. From these formulas, we can derive that C_{ext} is larger than C_{int} for all ratios between the drive and load transistor, when proper layout is taken into account. Therefore, this inverter family does not suffer from the self-loading effect.

$$\begin{cases} C_{int} = 2C_{gd2} + C_{gs1} \\ C_{ext} = C_{g4} + C_{gd4} = C_{gso4} + C_{gsc4} + C_{gd4} \end{cases} \qquad (3.41)$$

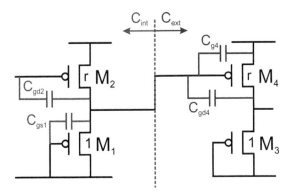

Figure 3.33   Intrinsic and extrinsic capacitors of a diode-load inverter.

C_{gd2} should be included more than once to take the Miller effect into account. Similarly as for the zero-V_{GS}-load logic, we have opted to use a factor of 2. From these equations, we can derive that C_{ext} is larger than C_{int} when the ratio r becomes larger than or equal to 1 and when proper layout of the transistors is taken into account. An asymmetrical layout results in a smaller C_{gd} versus C_{gs} for the drive transistor and a larger C_{gd} compared to C_{gs} for the load transistor.

> The diode-load inverter does not suffer from the self-loading effect, or in other words, the extrinsic capacitance is dominant when proper layout is taken into account. Sizing of these inverters will help to optimize delays!

The technology factor for diode-load inverters is derived in Eq. (3.42):

$$\gamma = \frac{2C_{gd2} + C_{gs1}}{C_{gso2} + C_{gsc2} + C_{gd2}} \tag{3.42}$$

γ will be smaller than 1, estimated to be approximately 0.68 for our calculated case having W/L transistor drive/load ratio of 10:1 and typical layout described in previous chapter, Section 2.3. The capacitances are listed in Table 3.4. The calculated γ suggests that this logic type is very useful for speed optimization. We can extract f_{opt} to be 3.33 from Figure 3.20.

Equation 3.32 shows the calculated minimum delay for a whole inverter chain. Figure 3.34 plots the normalized delay as a function of the number of inverter stages for different overall effective fanouts. In this technology, dedicated sizing of the inverters for different effective fanouts can be used to yield minimal normalized delays.

Table 3.5 discusses the normalized optimum delay in this technology. We can conclude that sizing is indeed very useful in this technology, demonstrated by the remarkable difference between the unbuffered, two stages and optimal number of inverter chains.

Table 3.4. Calculated and measured γ factor for diode-load inverter with the following dimensions: W/L equals 140/5 and 1400/5, designed in the technology described in Chapter 2, Sections 2.4.1 and 2.4.2; the channel capacitances are calculated for transistors operating in saturation mode

Ratio	A_{gd2} (μm^2)	A_{gs2} (μm^2)	A_{gsc2} (μm^2)	A_{gs1} (μm^2)	A_{gsc1} (μm^2)	γ
10:1	4285	5495	4620	725	462	0.68

Table 3.5. $\dfrac{t_p}{t_{d0}}$ for various effective fanouts

r	γ	f_{opt}	F	Unbuffered	Two stages	N_{opt}	Optimal chain
10	**0.68**	**3.33**	10	16.18	11.30	2 ~ round (1.91)	11.30
			100	148.53	31.41	4 ~ round (3.83)	22.60
			1000	1472.06	95.01	6 ~ round (5.74)	33.90

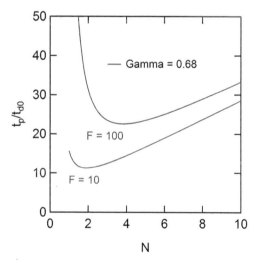

Figure 3.34 $\dfrac{t_p}{t_{d0}}$ plotted as a function of the number of inverter stages of overall effective fanouts and for γ of 0.68.

3.3.4 Dual V_T, Diode-Load Logic in Dual-Gate Technologies

Diode-load logic yields very good dynamic properties compared to zero-V_{GS}-load logic. The main concern with this logic family is the static behavior with low gains and corresponding small noise margins. Dual-gate, diode-load logic can be envisaged in improved static properties. Similarly to Section 3.3.2, we have realized dual-gate diode-load inverters whereby each transistor is exchanged with its

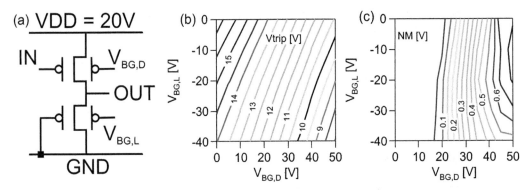

Figure 3.35   Scheme of a dual-gate diode-load inverter (a), contour plots of the trip voltage (b), and the noise margin (c) are determined from measured transfer curves of inverters when varying the back-gate voltage of the load and drive transistor and applying a supply voltage of 20 V.

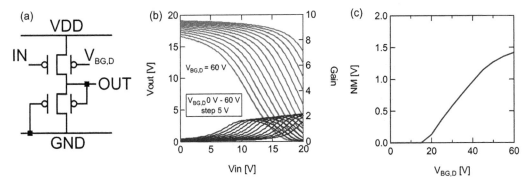

Figure 3.36   (a) Scheme of the optimized dual-gate diode-load inverter, (b) the corresponding measured VTC while sweeping the back-gate voltage of the drive transistor at a *VDD* of 20 V, and (c) the extracted noise margin as a function of $V_{BG,D}$.

dual-gate representative, as shown in Figure 3.35 (a). Figure 3.35 (b) and (c) plot the trip voltage and noise margin derived from the inverter's measured VTC when varying the back-gate voltages of the load and drive transistor. It can be seen that V_M can be shifted toward *VDD*/2 (here 10 V) and the obtained noise margins are small; however, the noise margins can be increased by increasing the back-gate voltage of the drive transistor. Moreover, increasing the back-gate voltage of the load transistor has less impact on V_M compared to the back-gate voltage of the drive transistor, which corresponds to Eq. (3.38).

In order to improve the noise margin, we have connected the back gate of the load transistor to the output node, resulting in the optimized dual-gate diode-load inverter configuration, which is based on the same effect shown previously with the optimized dual-gate zero-V_{GS}-load inverter. Figure 3.36 shows the scheme of the optimized dual-gate diode-load inverter and its measured VTC, gain, and noise

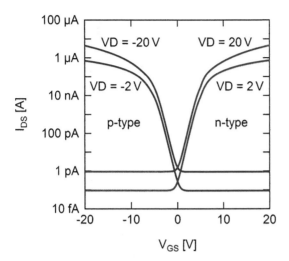

Figure 3.37 Transfer curve (linear and saturation) of a p-type and n-type thin-film transistor.

margin. The back-gate voltage of the drive transistor can be used to shift the VTC toward $VDD/2$. The extracted noise margin is now close to 1.5 V. The diode-load inverter configuration yields good dynamic behavior (see 3.3.3.3), but the addition of a back gate helps to improve the static properties of this logic family.

3.4 Complementary Logic

Complementary logic is the most powerful option to increase the noise margin of inverters and other logic gates, to have larger gains and faster delays. The large robustness of this topology against parameter variations leads to an increased yield compared to unipolar circuits. This caused a major revolution in circuit design some four decades ago and enabled technology scaling and the increase of complexity of Si-circuits. Therefore, we also expect a large impact in the field of thin-film circuits when a robust complementary technology is available. The schematic of a complementary inverter is depicted in Figure 3.1 (a). It consists of a p-type pull-up transistor and an n-type pull-down transistor. In contrast to unipolar inverters, both transistors actively drive the output node. At a low input voltage, the pull-up transistor is switched on, while the pull-down transistor is (almost) off, depending on threshold voltage. This pulls the output node to VDD. When the input voltage is high, the pull-down transistor pulls the output to 0 V while the pull-up transistor is (almost) switched off.

3.4.1 VTC of the Complementary Inverter

Figure 3.37 plots the transfer curves of both the n-type and p-type thin-film transistors as used in these simulations. Both have an ideal on-set voltage at 0 V.

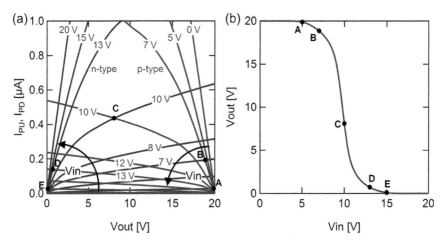

Figure 3.38   (a) Load curves of the p-type and n-type transistor and subsequent extraction of the inverter transfer curve for a *W/L* ratio between p-TFT:n-TFT of 1:1 and (b) corresponding inverter VTC.

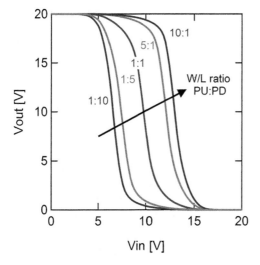

Figure 3.39   Influence of the *W/L* transistor ratio between the drive and load transistor on the VTC of the complementary inverter.

Similarly to that described previously, we can extract the inverter VTC from the load curves of the pull-up and pull-down transistors to find for each input voltage the DC operating point. Figure 3.38 shows the load curves and the corresponding extracted inverter VTC. The *W/L* transistor ratio is selected to be 1:1 (*PU:PD*).

The influence of the *W/L* ratio between the p-type and n-type transistor is shown in Figure 3.39. It can be observed that, if the p-type transistor is made larger, the trip point of the inverter moves to the right, since the p-type transistor is now stronger than the n-type. The same holds when increasing the ratio beneficial for the n-type transistor, which will lead to a shift of the inverter VTC to the left.

3.4.2 Static Behavior of the Complementary Inverter

3.4.2.1 V_M

The trip point is defined at the VTC where V_{in} equals V_{out}. We derive the equation for V_M similarly to the derivation for the zero-V_{GS}-load and the diode-load inverter, by describing the pull-up and pull-down currents.

$$\frac{1}{2}\mu_p C_{ox}\left(\frac{W}{L}\right)_p\left(V_{SG_p}+V_{T,p}\right)^2 = \frac{1}{2}\mu_n C_{ox}\left(\frac{W}{L}\right)_n\left(V_{GS_n}-V_{T,n}\right)^2 \qquad (3.43)$$

In this equation, we can assume that C_{ox} is equal for both the p-type and n-type transistors. The mobility values and threshold voltages are not matched in current state-of-the-art complementary technologies. The following equation also introduces V_M:

$$\mu_p\left(\frac{W}{L}\right)_p\left(VDD-V_M+V_{T,p}\right)^2 = \mu_n\left(\frac{W}{L}\right)_n\left(V_M-V_{T,n}\right)^2 \qquad (3.44)$$

Solving (3.44) leads to the equation for V_M for a complementary inverter

$$V_M = \frac{\sqrt{r}\left(VDD+V_{T,p}\right)+V_{T,n}}{1+\sqrt{r}} \quad whereby\ r = \frac{\mu_p\left(W/L\right)_p}{\mu_n\left(W/L\right)_n} \qquad (3.45)$$

We include in the ratio r also the difference in mobility between p-type and n-type transistors because they are not matched in thin-film transistors. From Eq. (3.45), we can also state that the ratio is the main parameter to influence the trip point for complementary inverters. The influence of the ratio on the switching threshold is shown in Figure 3.40. The ratio is plotted on a log-scale, to demonstrate the influence of re-sizing both n-type and p-type transistor. The influence of the ratio is in agreement with simulation results plotted in Figure 3.39. Moreover, the influence of the combination of a depletion device with an enhancement device is shown in this figure. From Equation. (3.45), we can also calculate that relatively small influences on the threshold voltage have only limited impact on the trip point.

3.4.2.2 Gain

The gain at V_M can be derived from the small signal analysis of the inverter. The small signal equivalent of the complementary inverter is shown in Figure 3.41. From this figure, we can derive the small signal gain for this inverter type in Eq. (3.46).

$$A_{V_M} = \frac{v_{out}}{v_{in}} = -\frac{g_{mp}+g_{mn}}{g_{op}+g_{on}} \qquad (3.46)$$

This formula indicates that the gain at V_M is larger than unity.

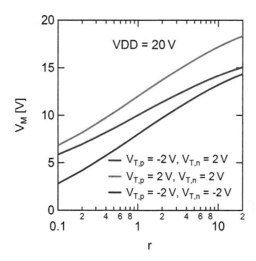

Figure 3.40   Influence of the transistor ratio (defined in Eq. (3.45)) between the p-type and n-type transistor on the switching threshold for enhancement devices with equal threshold voltage and for the combination of enhancement and depletion devices.

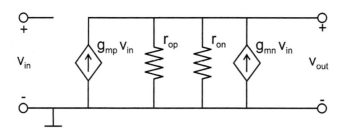

Figure 3.41   Small signal circuit equivalent of the complementary inverter.

3.4.3   Dynamic Behavior of the Complementary Inverter

3.4.3.1   Study of the Complementary Inverter Capacitances

$$\begin{cases} C_{int} = 2\left(C_{gd1} + C_{gd2}\right) \\ C_{ext} = C_{g3} + C_{g4} + C_{gd3} + C_{gd4} \end{cases} \tag{3.47}$$

The intrinsic and extrinsic capacitors of the complementary inverter are shown in Figure 3.42. We can derive that C_{ext} is larger than C_{int} for all ratios between p-type and n-type transistors. $C_{gd1,2}$ should be included more than once to include the Miller effect. Similarly as for other topologies, we have chosen a factor of 2.

> Also the complementary inverter does not suffer from the self-loading effect, or in other words, the extrinsic capacitance is mostly dominant. Sizing of these inverters will shorten delays!

Table 3.6. Calculated and measured γ factor for complementary inverters with the following dimensions: minimal W/L is 140/5; the technology to calculate the capacitances is based on Chapter 2, Section 2.4.6; the channel capacitances are calculated for transistors operating in saturation mode

Ratio (W/L) $PU:PD$	A_{gd2} (μm^2)	A_{gs2} (μm^2)	A_{gsc2} (μm^2)	A_{gd1} (μm^2)	A_{gs1} (μm^2)	A_{gsc1} (μm^2)	γ	$\gamma_{measured}$
1:1	537.5	2700	462	537.5	2700	462	0.29	No data
1:2	537.5	2700	462	965	3265	924	0.34	0.27

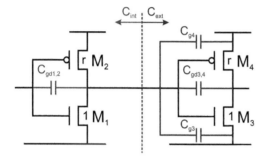

Figure 3.42 Intrinsic and extrinsic capacitors of a complementary inverter.

The technology factor for complementary inverters is derived in Eq. (3.48):

$$\gamma = \frac{2\left(C_{gd1} + C_{gd2}\right)}{C_{g1} + C_{g2} + C_{gd1} + C_{gd2}} \qquad (3.48)$$

γ will always be smaller than 1, estimated to be approximately 0.29 and 0.34 for our calculated case having a transistor size ratio of 1:1 and 1:2, respectively, designed using our typical TFT layout described in the previous chapter, Section 2.3. The baseline technology for these calculations is discussed in Section 2.4.6. The theoretical capacitances are listed in Table 3.6. For the layout, one has to take care to minimize C_{gd} of both transistors. The calculated value of γ implies that this logic type is very useful for stage delay optimization. We can extract f_{opt} to be 3.00 ($\gamma = 0.29$) and 3.04 ($\gamma = 0.34$) from Figure 3.20.

Equation 3.32 shows the calculated minimum delay for a whole inverter chain. Figure 3.43 plots the normalized delay as a function of the number of inverter stages for different overall effective fanouts. It is noteworthy that also in this technology (as in the diode-connected inverter), dedicated sizing of the inverters for different effective fanouts can be used to yield minimal normalized delays.

Moreover, we have realized the circuit described in Figure 3.17 (a) to predict the technology factor γ and the intrinsic delay t_{d0} from Eq. (3.26). Figure 3.44 (a) plots the obtained measured delay as a function of the increased load for transistor ratio

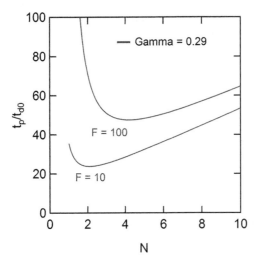

Figure 3.43 $\frac{t_p}{t_{d0}}$ plotted as a function of the number of inverter stages of overall effective fanouts and for γ of 0.29.

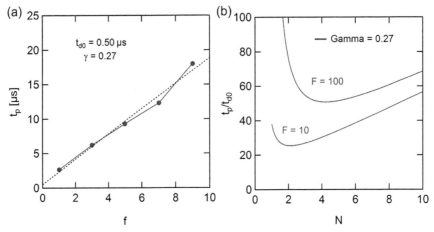

Figure 3.44 (a) Obtained measurements of a complementary inverter $(W/L)_p$:$(W/L)_n$ 1:2 with different loads and its fitted curve (black dotted line) and (b) $\frac{t_p}{t_{d0}}$ plotted as a function of the number of inverter stages for measured γ and overall effective fanouts.

1:2. The black dotted line is the linear fit which corresponds well with the measured data. The intrinsic delay can be extracted to be 0.50μs and γ is 0.27. The measured technology factor matches to the theoretical value listed in Table 3.6. Figure 3.44 (b) plots the normalized delay as a function of number of inverter stages for different overall effective fanouts.

Table 3.7. $\dfrac{t_p}{t_{d0}}$ for various effective fanouts

r	γ	f_{opt}	F	Unbuffered	2 stages	N_{opt}	Optimal chain
1:1	0.29	3.00	10	37.93	23.81	2 ~ round (2.10)	23.81
			100	348.28	70.97	4 ~ round (4.19)	47.62
			1000	3451.72	220.09	6 ~ round (6.29)	71.43
1:2	0.34	3.04	10	32.35	20.60	2 ~ round (2.07)	20.60
			100	297.06	60.82	4 ~ round (4.14)	41.20
			1000	2944.12	188.02	6 ~ round (6.21)	61.80
1:2 (*meas*)	0.27	2.98	10	40.74	25.42	2 ~ round (2.11)	25.42
			100	374.07	76.07	4 ~ round (4.22)	50.85
			1000	3707.41	236.24	6 ~ round (6.33)	76.27

Table 3.7 discusses the normalized optimum delay in this technology. We can conclude that sizing is indeed very useful in this technology, attested by the remarkable difference between the unbuffered, 2 stages and optimal number of inverter chains.

3.5   Conclusions

Table 3.8 summarizes the different logic families that we discussed using a variety of circuit metrics. We can conclude that both the zero-V_{GS}-load and diode-load inverter have an asymmetric trip point on the inverter VTC, assuming a single-V_T unipolar technology. On the contrary, complementary logic can be designed to have a trip point at half the supply voltage. The gain at the trip point is the largest for complementary inverters and the smallest for a diode-load connected inverter. Resulting noise margins will be smallest for diode-load inverters and largest for complementary inverters (higher gain compared to switching voltage around $VDD/2$). Keeping this in mind, from a static point of view, a complementary technology will yield the most robust inverters and therefore the possibility to integrate complex circuits with yield. The other side of the medal is that the necessity to have a p-type and n-type transistor integrated in a complementary technology is challenging in thin-film transistor technologies. Therefore, it is also shown that a unipolar technology having two threshold voltages can be a solution to increase the transistor's static parameters. A valuable option to have two threshold voltages available is by the addition of an individual back gate to each transistor. This showed improved static properties for zero-V_{GS}-load and diode-load families.

Table 3.8. Overview of circuit parameters for different logic families

	Zero-V_{GS}-load		Diode-load	Complementary	
r	$\dfrac{\left(W/L\right)_{PD}}{\left(W/L\right)_{PU}}$		$\dfrac{\left(W/L\right)_{PU}}{\left(W/L\right)_{PD}}$	$\dfrac{\mu_p\left(W/L\right)_p}{\mu_n\left(W/L\right)_n}$	
V_M	$VDD-\sqrt{r}\left(V_{T,PD}\right)+V_{T,PU}$		$\dfrac{VDD+V_{T,PU}-\sqrt{\dfrac{1}{r}}\left(V_{T,PD}\right)}{1+\sqrt{\dfrac{1}{r}}}$	$\dfrac{\sqrt{r}\left(VDD+V_{T,p}\right)+V_{T,n}}{1+\sqrt{r}}$	
A_{V_M}	$-\dfrac{g_{md}}{g_{od}+g_{ol}}$		$-\dfrac{g_{md}}{g_{od}+g_{ol}+g_{ml}}$	$-\dfrac{g_{mp}+g_{mn}}{g_{op}+g_{on}}$	
C_{int}	$2C_{gd2}+C_{gd1}$		$2C_{gd2}+C_{gs1}$	$2\left(C_{gd1}+C_{gd2}\right)$	
γ	$\dfrac{2C_{gd2}+C_{gd1}}{C_{gd2}+C_{gso2}+C_{gsc2}}$		$\dfrac{2C_{gd2}+C_{gs1}}{C_{gso2}+C_{gsc2}+C_{gd2}}$	$\dfrac{2\left(C_{gd1}+C_{gd2}\right)}{C_{g1}+C_{g2}+C_{gd1}+C_{gd2}}$	
γ	0.56 (r 1)	1.97 (r 10)	0.68 (r 10)	0.29 (1:1)	0.34 (1:2)
f_{opt}	3.23 (r 1)	4.30 (r 10)	3.33 (r 10)	3.00 (1:1)	3.04 (1:2)

From dynamic perspective, both complementary and diode-load inverters are relevant integrated circuit families to optimize inverter stage delays and the speed of logic chains. This is because the intrinsic capacitor is smaller than the extrinsic capacitor for the valid transistor ratios. Single-V_T, zero-V_{GS}-load inverters, on the other hand, suffer from the self-loading effect (C_{int} is larger than C_{ext}), whereby it becomes less advantageous to optimize the delay for a certain load. When the ratio of this inverter type can be designed 1 (e.g. two V_T's), the technology factor γ becomes lower than 1. In this case, optimization of chain delays makes again more remarkable differences. The introduced dual-gate zero-V_{GS}-load family can therefore exhibit γ lower to 1, which makes it a very relevant logic type to optimize delay paths and therefore a valid option for circuit integration.

In addition, we compare in Chapter 5 (Section 5.4.3) the speed performance for these logic families. This data has been obtained from measurements on the same substrate. Zero-V_{GS}-load logic results in the largest value for stage delay when extracted from 19-stage ring oscillators. The main reason is the limited current available from the load transistor, compared to diode-load and complementary logic. Latter named logic families yield similar stage delays.

3.6 Suggestions to Improve the Inverter Performance

3.6.1 Level-Shifter

In order to be able to integrate more logic gates in a circuit, the use of level-shifters has shown to be successful as it improves the noise margin of unipolar inverter types [142], [125]. A level-shifter adds an additional stage to the inverter and requires a third power rail. This circuit results in a shift of the transfer characteristics of the inverter, whereby the trip point and the noise margin can be improved. This is attested by the demonstration of relatively large circuits with this configuration [94]. The disadvantage of this solution is that each inverter requires an additional stage, which will increase the area of the circuit and therefore affect the yield, when it is limited by hard faults.

3.6.2 Self-Aligned Technology

One technological issue in organic and metal-oxide thin-film transistor technologies is the fact that source-drain contacts are fully overlapping the gate metal. Therefore, there is always inherently a substantial parasitic gate-source and gate-drain capacitance present. This will increase the intrinsic and extrinsic capacitors of the logic types and the corresponding intrinsic stage delay. One method to reduce the parasitic overlap capacitances strongly is to move toward a self-aligned technology, which results in zero (or minimal) overlap between source-drain electrodes and gate. Such technologies will be very beneficial for the stage delay. It is also pursued for TFTs in the active matrix of displays, as the minimum overlap capacitance results in a faster RC time constant that determines the write speed of data in display pixels. Sony has recently demonstrated an AMOLED display based on a self-aligned, metal-oxide transistor backplane [65].

For all described topologies, the proportional factor γ could be lowered using this method by reducing the parasitic overlap capacitors. The only topology that will not profit from this benefit is the zero-V_{GS}-load inverter with large ratios (e.g., 10), because C_{ch} of the load transistor is present in the intrinsic capacitor and not in the extrinsic, nor in the input capacitor.

4 Variability

Process variations present during transistor fabrication lead to a certain variability in the resulting transistor parameters. Current and upcoming process nodes for Si-CMOS technologies are such an example where parameter variability increases substantially at subsequent nodes as a result of downscaling of the device dimensions toward fundamental limits of the technology [143]–[147]. Another field that suffers from parameter variability is the field of ultra-low power circuit design based on subthreshold logic, that is, logic in which transistors are used in the sub-threshold regime [148]–[151]. Considerable parameter variability is also present for organic and metal-oxide thin-film transistors (on foil). Different categories of process variations will be discussed first in this chapter, followed by a qualitative listing of the sources of process variation. Next, the consequences of such process variations on circuit yield are analyzed. Finally, designs can cope with these process variations. This will be investigated in more detail for short-range and long-range non-uniformities.

4.1 Classifications

Four categories for process variations on transistor parameters can be defined:

- Batch-to-batch: deviations of the parameters of devices between different batches,
- Wafer-to-wafer: deviations of the parameters of devices between different wafers in the same batch,
- Die-to-die or inter-die (D2D): deviations of the parameters of devices on separate chips on the same wafer,
- Within-die or intra-die (WID): deviations of the parameters of devices on the same chips.

Multiple sources of variation for organic and metal-oxide thin-film transistors will be discussed in the next section. These variations find their origin in materials or technology and have an impact on one or more parameters of the saturation current equation of the thin-film transistors (Eq. (4.1)).

$$I_{DS,sat} = \frac{1}{2}\mu C_{ox}\frac{W}{L}\left(V_{GS}-V_T\right)^2 = \frac{1}{2}\mu\frac{\varepsilon_r\varepsilon_0}{t_{ox}}\frac{W}{L}\left(V_{GS}-V_T\right)^2 \qquad (4.1)$$

In this equation, μ is the charge carrier mobility. The thickness and the relative permittivity of the dielectric are given by, respectively, t_{ox} and ε_r. Another important parameter is the threshold voltage, noted by V_T. W and L are, respectively, the width and length of the transistor's channel. From this equation it is clear that variations in the threshold voltage will affect the saturation current of the transistor quadratically, while variations in μ, C_{ox}, and device geometry influence the saturation current only linearly.

From a dynamic point of view, variations in the geometric sizes of transistor capacitances also count. The previous chapter discussed the dynamic impact of these capacitances on the circuit behavior.

4.2 Sources of Process Variation

In this section, sources of process variation will be discussed in a qualitative way for organic and metal-oxide thin-film transistor technologies. The sources are categorized by their origin in the semiconductor, dielectric, contacts, and foil.

4.2.1 Semiconductor

There are many sources of polycrystalline process variations for organic and metal-oxide thin-film transistor technologies (on foil). Organic semiconductors, such as pentacene, form poly-crystalline layers with randomly distributed grain boundaries and grain sizes [152]. The presence of grain boundaries introduces broad tail states [153] and makes the semiconductor mobility much more dependent on, among others, the electrical field [154]. This affects the mobility and threshold voltage [155] of the resulting transistors. It can be presumed that longer channel lengths (L) and/or larger devices (W) should lower the device parameter variation because of the averaging of these grain boundaries. Low-temperature, metal-oxide semiconductors and amorphous organic semiconductors, on the contrary, have an amorphous nature, which may be more beneficial with respect to parameter variations.

In both cases, the deposition technique and deposition conditions will also have an influence on the device performance. These deposition techniques can vary from solution-processing (e.g., spin coating [94] or ink jet printing [156]) to vacuum coating techniques (evaporation, sputtering, organic vapor phase deposition [27], and so on). During deposition of the semiconductor, the growth of the first monolayers is of paramount importance for the transport at the interface. Also the surface energy [152] – which can be controlled by proper cleaning and/or surface treatments – and the dielectric roughness [157] play a role in the growth of the semiconductor. All process variations affecting the parameters mentioned influence device performance.

In the process flow toward an integrated transistor, the chemical robustness and physical robustness of the semiconductors against all subsequent steps in the process flow are also critical. These steps could affect the device performance: either improve it in terms of, for example, air stability or improved bias stress or degrade the performance, as is often experienced by a shift in threshold voltage or a decreased mobility. As a final note, organic semiconductors are often subject to randomly distributed unintentional doping leading to substantial threshold variations [158].

4.2.1.1   Dielectric

The dielectric is, besides the semiconductor, the most critical layer in the technology. There are a lot of requirements to form a good dielectric:

- High breakdown voltage,
- Low leakage current,
- Well-controlled thickness (influences local C_{ox}),
- Low roughness level [157],
- Well-controlled density (influences ε_r),
- No charge traps, either in the bulk or at the interfaces [159],
- Chemical and physical resistance against processing of the subsequent layers: contacts, semiconductor, and so forth.

Process variations on the dielectric thickness and density (C_{ox}) will mostly have an impact on the D2D characteristics of the devices, more specifically on the threshold voltage and even the mobility in the case of field-dependent mobilities. The variation in thickness of the dielectric is dependent on the technology and tools used. Solution coating and chemical vapor deposition are two industrial techniques. Lab techniques include sputtering, thermal evaporation, and atomic layer deposition (ALD). The roughness level may be different for devices in the same die. Impact of roughness level on charge carrier mobilities for organic thin-film transistors has been studied already [157]. This local roughness can stem from roughness variations of the underlying layers such as the gate metal. Charge traps, influencing the threshold voltage, mostly contribute to WID variations. These can be present in the bulk of the dielectric material or at the interface between dielectric and semiconductor. The subsequent steps in the fabrication process toward an integrated transistor may also influence the quality of the gate dielectric. Plasma etch steps are a notorious source of V_T variation and can be limited by reducing the antenna ratio [160]–[162].

4.2.2   Contacts

A variety of deposition techniques can be used to define gate and source/drain contacts. Industrially, sputtering is mostly used. In particular for organic TFTs, exploratory techniques such as evaporation, solution processing, and printing have been demonstrated. Besides the process variation with respect to the deposition technique, the patterning will also have an impact on process and device variation. It will influence the shape of the contacts and eventually the line edge roughness of the sources and drains. Line edge roughness gives rise to variations on the local channel length of a device [163]–[165], influencing the source-drain field. Longer channel lengths (L) and/or larger devices (W) may lower the device parameter variation because of the averaging of line edge roughness effects. Moreover, variations on the contact resistance of source and drain, which impacts the charge carrier injection into the semiconductor, might provide additional variations on the device performance [19], [166]–[171]. The influence of contact resistance is larger for shorter channel lengths; hence this effect is rising in importance when channel length downscaling is required, for example, to increase the circuit's speed. Contact resistance is also dependent on the device cross section. Coplanar devices have a larger contact resistance compared to the staggered architectures. The larger area over which charge carrier injection occurs in the staggered device explains the lower contact resistance with respect to coplanar devices. Moreover, the source-drain contacts also shield the gate field in a coplanar device.

4.2.3 Foil

A key feature for thin-film transistor technologies is the possibility to process directly on foil, which could lead to potential new applications, among others flexible displays and flexible circuitry. Although applications on flexible substrates may be beneficial for the end-user, processing-wise foil has its drawbacks. One important disadvantage is the limited temperature budget available during processing. It prevents the use of annealing, often employed to improve stability of the semiconductor, and limits the possible deposition techniques for dielectrics, leading to lower quality dielectric layers. Moreover, during processing on foil (by roll-to-roll coating or the foil-on-carrier approach), the dimensional stability of the foils during the subsequent process steps is not guaranteed [172], resulting in larger D2D process variations.

We should note, finally, that there is an important difference between fabs and labs. Tools and equipment for labs are usually shared for research and development, therefore inherently leading to more process variations. Fabs, on the contrary, have dedicated tools and are reluctant to make any process flow modifications. It is clear that this leads to better process control, especially for wafer-to-wafer and batch-to-batch variations. Many reports based on organic and metal-oxide thin-film transistors are performed in laboratories.

4.3 Influence of Parameter Variation on the Yield of Logic Circuits

In Chapter 3, the noise margin was discussed as a useful figure-of-merit to determine the yield of complex logic circuits that integrate a large number of logic gates with unavoidable parameter spread (e.g., on V_T and/or mobility). Process variation and consequently transistor parameter variation will have an influence on the performance and yield of logic gates and circuitry. In this section, the influence of

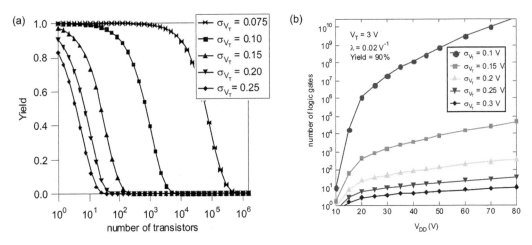

Figure 4.1   (a) Worst-case calculation of the circuit yield as a function of the number of zero-V_{GS}-load inverters, calculated for $VDD = 15$ V and an average V_T of 3 V; (b) the average noise margin increases for increasing VDD, such that the number of logic gates that can be integrated also increases (other parameters kept the same as for (a)).

WID variations of the threshold voltage on the static yield of circuits will be discussed. The typical spread on threshold voltage in organic TFT technology is 0.2 to 0.5 V.

De Vusser et al. have calculated for a worst-case scenario the yield as a function of the number of concatenated unipolar zero-V_{GS}-load inverter stages with a driver/load (W/L) ratio of 1:5 [137]. The yield is calculated as the probability that the series connection of all inverter stages will have a noise margin larger than 0. For these calculations, the average V_T is assumed to be 3 V and the circuit's supply voltage equals 15 V. Figure 4.1 plots the result of the yield calculations. The right panel shows that the number of stages that can be integrated for a 90 percent yield increases with VDD thanks to the increasing average noise margin. This explains why many organic unipolar circuits with a significant amount of transistors require substantial VDD to operate properly.

As discussed in Chapter 3, complementary logic has clear advantages in terms of static performance compared to unipolar logic, among others a larger noise margin. Similar to the calculations for zero-V_{GS}-load logic, Bode et al. have calculated the yield as a function of concatenated inverter stages based on complementary inverters and compared it to zero-V_{GS}-load logic [173]. Figure 4.2 (from [174]) plots this comparison for an average V_T of 2.5 V with a supply voltage of 15 V. The yield analysis for complementary logic predicts a yield loss of only 0.6 percent in comparison with a yield loss of practically 100 percent for unipolar p-type, zero-V_{GS}-load logic for a circuit complexity of 10^4 equivalent inverters and a variation on V_T of 0.5 V. In other terms, complementary logic is more forgiving in terms of static yield loss because of parameter variations than unipolar, single-V_T, zero-V_{GS}-load logic.

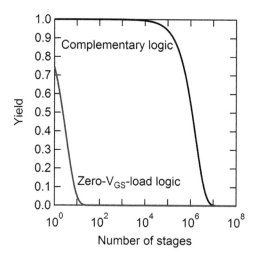

Figure 4.2 Comparison of the predicted yield for unipolar and complementary circuits having a standard variation on V_T of 0.5 V [174].

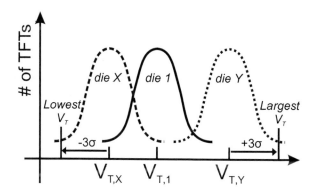

Figure 4.3 Schematic representation of multiple WID variations of the threshold voltage over the same wafer.

4.4 How to Cope with WID and D2D Parameter Variations

When the targeted circuit complexity exceeds inverters and ring oscillators, typical device parameters and their deviations should be modeled in order to be able to cope with these variations during the design phase. Figure 4.3 plots schematically an example of WID variations of the threshold voltage over multiple dies on a single wafer. Each die will have an average value and spread on the threshold voltage. The parameter variability does not need to be constant for every die on the wafer. The minimum and maximum threshold voltages on the wafer are important when performing D2D analysis, as will be explained further.

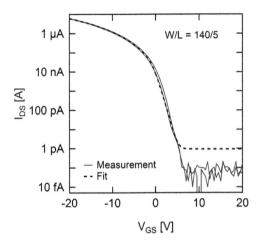

Figure 4.4   Transfer characteristics of an organic transistor on foil operated in saturation regime, with *W*/*L* of 140/5. The full line shows the mean curve obtained, while the dashed line represents the corresponding fit.

4.4.1   Designing with WID Variations

WID variations can be analyzed by means of Monte Carlo (MC) circuit simulations [175]. MC simulations will perform a number of subsequent circuit simulations, whereby, for each simulation run, the selected parameters (such as V_T) are distributed randomly across all transistors according to predefined mean and spread values. By performing a multitude of MC runs, the yield of the evaluated circuit can be estimated. This analysis tool is available for SPICE. As a starting point, it is very important to have detailed data on WID variations of the targeted technology. The technology used for the analysis in this chapter is the baseline technology for most of the unipolar circuit realizations discussed later in this work. The technology has been developed by Polymer Vision for commercialization in rollable active matrix displays and is described in Chapter 2, Section 2.4.1. Figure 4.4 depicts the transfer characteristics operated in saturation regime, representing the mean value of V_T for this technology. The solid line shows the measured curve after TFT fabrication, and the dashed line represents the corresponding fit as generated from the circuit simulator.

Since this technology has been developed for display applications, the absolute value and the spread on on-state current at large negative V_{GS} are key parameters. Figure 4.5 depicts the ratio between standard deviation on I_{on} and average I_{on} versus the transistor area, obtained for this technology. This data are collected on a full 150 mm wafer; 132 different transistors have been measured for five different transistor sizes, namely, 7/5, 14/5, 28/5, 56/5, and 140/5. The typical layout of these TFTs is shown in Figure 4.5. The percentage of spread in I_{on} increases linearly by decreasing square root of transistor area, apart from the smallest transistor size (7/5). We assume that edge effects are dominant for the smallest TFT with a semiconductor island width of only 3.5 μm, compared to larger devices.

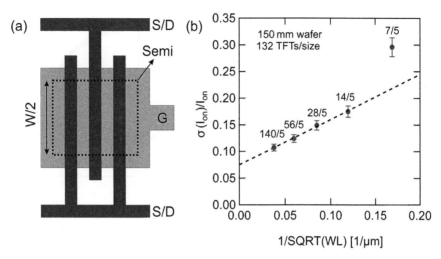

Figure 4.5 (a) Typical layout of these p-type organic TFTs and (b) standard deviation of the on-state current (I_{on}) plotted versus the inverse transistor area, measured on a 150 mm wafer. The dots are the measurements obtained; the dashed line is the corresponding fit.

A possible explanation for this effect could be averaging of process variations with increasing area, as explained in Section 4.2. For Si-CMOS technologies, Pelgrom et al. derived in 1989 a model for matching properties of MOS transistors [176]. They concluded that the variance of the threshold voltage and the current factor β, which is $\mu C_{ox} \dfrac{W}{L}$ between two adjacent devices, are inversely proportional to the transistor area.

The data in this book evaluate the variance for single devices, contrary to Pelgrom's model, which calculates the mismatch between two devices. The difference between approaches is explained in Eq. (4.2) using the threshold voltage as an example. The variance of the difference in threshold voltage for two devices (distributed randomly) is the sum of their individual variances. There is a factor 2 difference between approaches. In this work, for WID variations, we use the standard deviation of the absolute parameter.

$$\sigma^2\left(\Delta V_T\right) = \sigma^2\left(V_{T,TFT1}\right) + \sigma^2\left(V_{T,TFT1}\right) = 2\sigma^2\left(V_T\right) \qquad (4.2)$$

The dashed line in Figure 4.5 corresponds to a linear fit based on the measured data, leading to Eq. (4.3). The slope of the curve is found to be 0.85 ± 0.14 µm.

$$\frac{\sigma\left(I_{on}\right)}{\overline{I_{on}}} = \frac{0.85 \pm 0.14}{\sqrt{WL}} + \left(0.07 \pm 0.01\right) \qquad (4.3)$$

This data suggest that Pelgrom's mismatch law is valid for organic TFTs. Equation 4.3 also shows a non-zero variability for infinite area, suggesting that data collected

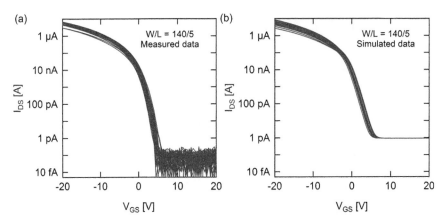

Figure 4.6   (a) Measured and (b) simulated WID variations of p-type organic TFTs with W/L 140/5, described in Section 2.4.1. The extracted standard deviation for V_T equals 0.35 V.

on the full 150 mm wafers suffer from D2D variation. The smallest TFT used for circuit realizations in this technology is 140/5, integrated as the drive TFT in a zero-V_{GS}-load logic gate. The data previously discussed are based on analysis for a 150 mm wafer. Applications such as low-cost, lower-end electronics constitute a smaller chip area. Hence, data on WID variations for smaller areas are required in order to predict circuit yield better. Figure 4.6 (a) shows an example of WID variations, obtained from 16 measured transistors with W/L of 140/5. These transistors are distributed with a pitch of 500 μm horizontally and 650 μm vertically. The extracted standard deviation for the threshold voltage is 0.35 V for a 140/5 transistor. Figure 4.6 (b) is the result of 40 MC simulation runs in the circuit simulator, including standard deviations on charge carrier mobility and threshold voltage.

Similar analysis has been made for a transistor with an increased geometry, namely, 1400/5. This is the chosen load transistor for our zero-V_{GS}-load inverters and NAND-gates. Figure 4.7 plots the measured and simulated WID variations of these transistors. The extracted standard deviation on the threshold voltage for the measured transistors is 0.13 V. When comparing Figure 4.6 to Figure 4.7, it is clear that the variation on the I_{DS}-V_{DS} characteristics is smaller for a larger device.

In terms of absolute numbers, the smaller device (140/5) exhibits a standard deviation on V_T of 0.35 V, while a 10-fold larger device yields only 0.13 V standard deviation. Figure 4.8 plots the standard deviation as a function of the square root of the inverse transistor area, both for extracted threshold voltage and for mobility. Equation (4.4) describes this dependence on threshold voltage and charge carrier mobility for closely spaced devices.

$$\begin{cases} \sigma(V_T) = \dfrac{A_{V_T}}{\sqrt{WL}} \\[2mm] \dfrac{\sigma(\mu)}{\mu} = \dfrac{A_\mu}{\sqrt{WL}} \end{cases} \qquad (4.4)$$

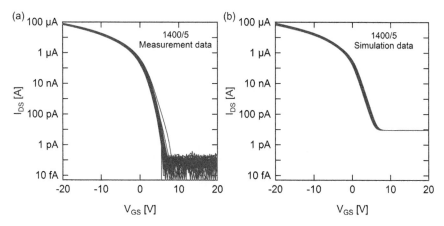

Figure 4.7   (a) Measured and (b) simulated WID variations of p-type organic TFTs with W/L of 1400/5. The extracted standard deviation for V_T equals 0.13 V.

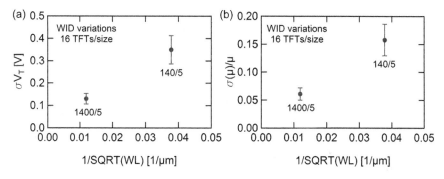

Figure 4.8   WID standard deviation of (a) the threshold voltage and (b) the mobility of p-type organic TFTs plotted versus the square root of the inverse transistor area for 140/5 and 1400/5 TFTs.

$\sigma(V_T)$ and $\sigma(\mu)$ are hereby defined as the standard deviation of the threshold voltage and the charge carrier mobility, respectively. $\overline{\mu}$ is the average mobility. A_{V_T} and A_{μ} represent both area proportionality constants, which will differ by a factor of $\sqrt{2}$ compared to the constants obtained for Pelgrom's mismatch model [176] according to Eq. (4.2). The proportionality constants in this work are extracted to be *9.9 ± 1.8 Vμm* for A_{V_T} and *4.5 ± 0.8 μm* for A_{μ}. The absolute values of the proportionality constants obtained may be used for MC simulations, but only with the necessary precautions because of the very limited number of data points available and consequently large error margin on A_{V_T} and A_{μ}. Future work on this topic should include more data points in order to obtain more accurate values for these proportionality factors.

Since the designs in this technology comprise only 140/5 and 1400/5 transistors, used as drive and load TFT, respectively, two different transistor models have been defined in the design library. Each transistor size has its own values

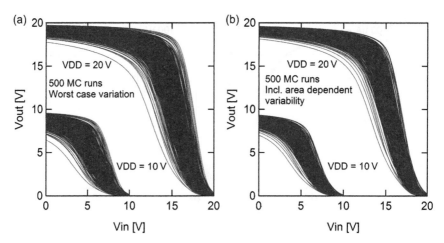

Figure 4.9   500 MC simulated zero-V_{GS}-load inverter VTC curves at 10 V and 20 V supply voltage by using (a) only the worst-case model for variation and (b) two different TFT models with area dependent variation.

for variation on charge carrier mobility and threshold voltage in the model file, extracted from measured data plotted in Figure 4.6 and Figure 4.7. Of course, one could already argue to use very large transistors in order to get rid of large variations and to make sure the circuit is fully operational. Increasing the size of the transistor will be at the expense of area, power, speed (especially when increasing the channel length), and last but not least circuit cost. In addition, larger transistor size increases the chance of non-working circuits due to particle related shorts and other type of hard faults. This trade-off should always be considered during the design phase.

Figure 4.9 shows the simulated VTC curves of zero-V_{GS}-load inverters with a drive/load transistor ratio of 1:10 for 500 MC runs. These simulations have been performed at 10 V and 20 V supply voltages. Figure 4.9 (a) plots the simulated VTC curves when using the worst-case values for standard deviation (140/5, Figure 4.6) for both TFTs. Figure 4.9 (b), on the other hand, includes the observations of decreasing spread by increasing transistor area. For the latter named simulations, parameters extracted from Figure 4.6 and Figure 4.7 have been used. Hence, the impact of having less parameter spread on the load TFT is visualized in Figure 4.9. These inverters exhibit less variation in the resulting VTC, which will lead to an improved yield prediction, especially at lower supply voltages.

As a next step, the predicted yield of more complex circuits has been analyzed using MC simulations. The 100 MC runs have been performed at different supply voltages for 19-stage ring oscillators in order to obtain an estimated yield prediction (see Table 4.1). This ring oscillator acts in the RFID transponder chip design as a clock; see Chapter 5. It is observed that increasing the supply voltage will increase the circuit yield, in agreement with Figure 4.1 (b). Moreover, simulations

Table 4.1. Simulated yield extracted from 100 MC simulated 19-stage ring oscillators for varying supply voltages between 7.5 V and 20 V

Yield of a 19-stage ring oscillator (%)	VDD					
	7.5 V	10 V	12.5 V	15 V	17.5 V	20 V
Worst-case variation	67	97	99	100	100	100
Including area dependent variability	92	99	100	100	100	100

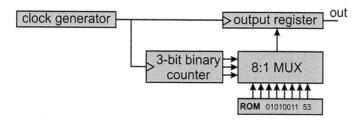

Figure 4.10   Circuit block diagram of an 8-bit RFID transponder chip.

taking into account the area dependent variability increase the predicted yield, as expected.

As a final step for WID variations, besides circuit simulations on all main logic gates and circuit building blocks, the same analysis should be done for the complete targeted circuit design, in this case, an 8-bit RFID transponder chip. The block diagram is shown in Figure 4.10. The yield predictions after 25 MC runs for this circuit are listed in Table 4.2. After 25 MC simulations, it turned out that the estimated yield at low supply voltages is unacceptably low, being 28 percent at 10 V supply voltage even when area dependent variability is taken into account. After analyzing the data, the clock signal seems to be a yield killer for a multitude of these runs and appears to be very weak to drive all clock-triggered flipflops in the scheme. This issue did not appear in the behavioral simulations excluding WID variations. After implementation of a buffer stage (buf3) just behind the ring oscillator, the yield becomes larger than 50 percent at 10 V supply voltage. Moreover, at 15 V supply voltage, the estimated yield is already 96 percent. Comparing in the end the estimated yield with and without area dependent variability shows big differences, especially at low supply voltages. This demonstrates the usefulness of having as many data as possible available prior to starting to design complex circuits in order to be able to optimize the circuit correctly. It is interesting to note that the observation of an increasing yield with supply voltage is also valid for the full circuit. This route to take WID variations into account has been used to design the transponder chips realized in the next chapters. The supply voltage at which all measured RFID transponder chips are fully operational is 14 V, which matches the analysis made in this section.

Table 4.2. Simulated yield extracted from 25 MC simulated 8-bit transponder chips for varying supply voltages between 10 V and 20 V

Yield of an 8-bit transponder chip (%)	VDD				
	10 V	12.5 V	15 V	17.5 V	20 V
Buf3 @ clk, worst-case variation	4	36	64	80	96
Area dependent variability, but no buf3 @ clk	28	68	88	96	96
Area dependent variability and buf3 @ clk	56	80	96	100	100

4.4.2   Designing with D2D Variations – Corner Analysis

D2D variations can be investigated during corner analysis studies. In this analysis, one explores whether the circuit still performs within the specifications at predefined parameter corners such as for speed specifications. It allows decisions to redesign or optimize the circuit when such yield loss in corners has been observed. The definition of these corners is schematically depicted in Figure 4.3. A full wafer contains a specific number of dies exhibiting a Gaussian distribution of the threshold voltage per die. In this particular case, dies X and Y show the lowest and largest threshold voltages. The minimum and maximum corners are defined at 3σ difference from the average V_T, in this case, $V_{T,X} - 3\sigma$ and $V_{T,Y} + 3\sigma$, respectively. As a general remark, corners are not only defined for threshold voltages, but may also be applied to the supply voltage, temperature, and so forth. Moreover, in this work, we have chosen 3σ to define the corners; technology providers usually take more margin in order to guarantee that all transistors comply with these margins.

When performing corner analysis, usually the following data are present: slow corner, fast corner, and typical corner. As an example, the slow corner for n-types is defined at the largest value of the threshold voltage. In this corner, the current that a transistor at a given die can provide is less compared to that of other dies. Hence, this corner is useful to analyze whether the circuit designed can operate at the targeted specifications with respect to speed requirements. Such corner analysis is also important to determine the logic type for unipolar circuits, listed in the previous chapter. As an example, the most reported unipolar logic gate design is based on zero-V_{GS}-load logic. This logic type requires a load transistor that operates in depletion mode. On the other hand, when the transistor behaves in enhancement mode (e.g., in the slow corner), this transistor may be switched off or may operate in subthreshold regime, yielding non-functioning circuits and/or very slow performing circuits.

Figure 4.11 visualizes a clear difference in corner definitions between unipolar and complementary technologies. There is only a 1D plane available in the case of unipolar technologies because of the presence of a single semiconductor type, depicted in Figure 4.11 (a). The threshold voltage of the resulting transistors can be typical (T) or can lead to fast (F) or slow (S) circuit operation. In complementary

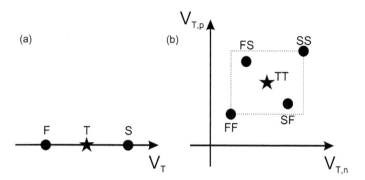

Figure 4.11   Corner definitions for (a) unipolar and (b) complementary technologies.

technologies, on the other hand, both n-type and p-type transistors are available. The threshold voltages of both TFTs can vary across the wafer, leading to typical-typical (TT), fast-fast (FF), and slow-slow (SS) corners, as indicated in Figure 4.11 (b). Moreover, combinations of fast and slow corners for complementary technologies are present, more specifically the FS and SF corners. Remarkably, the FS and SF corners are not located on the corners of the square drawn in Figure 4.11 (b), but inside the square described by the FF and SS corners. This can be explained because some process variations for both n-type and p-type transistors are correlated, such as those due to *global* variation of the dielectric thickness.

4.4.3   Adaptive Back-Gate Control for Threshold Voltage Compensation

Previous chapters described that thin-film transistor technologies can be fabricated with an additional metallization layer that can be used as an extra gate acting as a back gate or V_T-control gate. The main advantage of a back gate for thin-film transistors is that each individual thin-film transistor can have an individual back gate. Therefore, as explained in the previous chapter, unipolar, multiple threshold voltage circuits can be realized. In the next part of this section, we will discuss how adaptive back-gate control can be used not only to obtain increased robustness, but also to comply with speed specifications of the targeted circuit.

Figure 3.23 depicts the scheme of a p-type zero-V_{GS}-load inverter whereby each transistor is replaced by a dual-gate transistor. As a consequence, both transistors can be driven with different back-gate voltages, yielding a dual-V_T zero-V_{GS}-load inverter with improved robustness, as discussed in the previous chapter. Figure 3.23 (b) and (c) plot the impact of applying back-gate voltages to the drive and the load TFT independently, resulting in an improved static noise margin under correct biases. One of the disadvantages of this zero-V_{GS}-load topology is the necessity to have two extra voltage rails or internally generated voltages available to correct for variations. In the previous chapter, an optimized scheme of a dual-gate zero-V_{GS}-load inverter requiring only a back-gate voltage for the drive transistor was introduced. Moreover, this inverter configuration yielded larger noise margins,

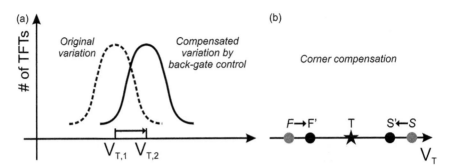

Figure 4.12   Schematic representation of (a) the impact of a shift in threshold voltage on WID variations and (b) corner compensation by applying a back-gate bias voltage.

as explained in the previous chapter. These excellent static properties can account for even larger variations, which enable integration into larger circuitry, such as RFID transponder chips [115] and the first organic 8-bit microprocessor [63] on foil. These realizations will be described in more detail in the following chapters.

With an increased noise margin compared to single-V_T zero-V_{GS}-load logic, dual-V_T zero-V_{GS}-load logic will have a beneficial impact on the integration density and yield of circuits with the same standard deviations. This is because the average threshold voltage of the die can be shifted in order to compensate for 3σ variations on V_T, schematically depicted in Figure 4.12 (a). This technique allows correction for WID variations on full circuit level, by a limited number of bias voltages, for example, to control all load and/or drive TFTs.

D2D variations can also be taken into account using adaptive back-gate control for threshold voltage compensation. Figure 4.12 (b) depicts schematically that the original fast (F) and slow (S) corners can be shifted closer to the typical values by adapting a back-gate voltage, resulting in F' and S'. An integrated solution for this method may be critical path replica tuning. Critical path replica tuning by adaptive back-gate bias or supply voltage is used for post-silicon delay prediction in current Si-CMOS realizations [177], [178]. As such, a replica of the critical path is designed in each die and the performance will be evaluated during operation. The back-gate voltage (or VDD) is corrected by an additional compensation circuit in order to comply with, for instance, targeted clock requirements. This compensation circuit will consequently apply the same back-gate voltage to the targeted circuit.

Since both threshold voltages can be adjusted by the back gate, the operational frequency of the targeted circuit can also be regulated. Figure 3.24 shows the measurement results of the extracted frequency of 19-stage ring oscillators when varying the back-gate voltages of the load and drive transistors. By adjusting both threshold voltages, the operational frequency can be tuned, in this case, between 3 kHz and 24 kHz. This compensation technique allows matching of the speed targets by using two voltage rails. Also here, critical path replica tuning may offer an integrated solution.

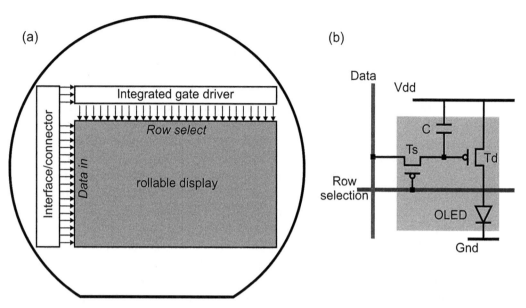

Figure 4.13 (a) Schematic representation of a flexible AMOLED display fabricated on wafer-level and (b) AMOLED pixel drive scheme based on two p-type transistors and one capacitor.

4.4.4 Variability and Large Area Electronics

One of the main future outlooks for the field of thin-film-transistor technologies on foil is the potential application field of large area electronics. Such applications could be large flexible displays [105], [94], [179] or sensor-distributed networks on large areas [60], [180], [181]. An example of a flexible AMOLED display is schematically represented in Figure 4.13.

The simplest pixel drive scheme for active matrix OLED displays is depicted in Figure 4.13 (b). An organic LED (OLED) is a current driven device that emits light, whereby a change in current through the OLED results in different brightness values for the OLED. In this pixel scheme configuration, the applied gate-source voltage of the drive transistor (T_D) regulates the drain-source current I_{DS} that flows through the OLED. This current is kept stable by the gate-source capacitor C. The pixel data values can be stored on this capacitor when the select transistor (T_S) is active. The drive transistor acts as a current source and is therefore mostly biased in saturation regime. This implies that the resulting current level is proportional to the mobility and the geometry of the transistor and quadratically proportional to V_{GS}-V_T. Variations on all these parameters will directly impact the brightness level of the OLED. Moreover, during operation, the V_T of both transistors can shift as a result of bias stress effects. As shown in Figure 4.13, considering only WID variations for the pixel scheme will not be effective, since the display takes almost the complete area of the wafer. In this case, the active area of the display becomes the die. If the level of these variations may cause observable non-uniformities in the brightness distribution, display designers consider means to overcome this issue by (limited) addition of transistors and capacitors to the pixel design in order to compensate, for example, for V_T variations at each pixel [9], [6], [32], [33], [182]–[184], [18], [185], [186]. Obviously, the addition of transistors and capacitors is limited by the display resolution and consequently the pixel size, but also by the transistor performance, which allows designing smaller transistors in the pixel engines.

Flexible displays require gate drivers fabricated in the same TFT technology on foil as the backplane, in order to reduce the number of wires necessary to drive this display [94], [105]. As Figure 4.13 depicts for gate driver application, the width of the line driver is equal to the total display width, while the height is preferably kept very limited in order not to lose too much area for the periphery and to maximize the active display area. The design and simulations of the line driver need to be done very carefully, since its width extends over almost the complete wafer. Therefore, simulations using only WID variations will not be sufficient. The in-depth knowledge of the technology's D2D variations is crucial for a dedicated line driver design. For such circuits, for which timing is very critical to refresh the display, simulations in the SS, SF, and FS (complementary) or S (unipolar) corners are necessary. These will influence design choices regarding operational speed while considering a reduction of the power consumption.

Another option to design the gate driver is to use the adaptive back-gate control option. The idea is to split the full length of the line driver into multiple sections (e.g., each section comprises 25 stages). The line driver design is optimized for one section using WID variations. D2D variations can be controlled by implementing one or two back-gate power rails per section (see Section 4.4.3), depending on the technology option chosen. During operation of the final circuit, these additional power rails can be used to tune the circuit to comply with speed specifications. Besides speed, the same technique of adaptive back-gate control can be used to improve the robustness of the circuit and therefore the final yield.

4.5 Conclusions

In this chapter, we have elaborated on the design challenges for mostly unipolar thin-film transistor technologies when process variability results in significant parameter variations, which limit the integration density of these transistors into circuits. Sources of these process variations in thin-film transistor technologies have been categorized by their origin in the semiconductor, dielectric, contacts, and foil. Furthermore, the use of a flexible substrate and the limited available temperature budget also contribute to process variability. Of course, the nature of the materials used in the transistor stack also plays an important role in parameter variation.

In the following discussion, the influence of parameter variation on the yield of logic circuits has been discussed, for both unipolar zero-V_{GS}-load logic and complementary logic. The latter logic type allows a larger integration density for the same

WID variation compared to zero-V_{GS}-load logic. During the simulation phase, WID variations can be taken into account by means of MC mismatch simulations. These simulations enable prediction of the yield of the targeted circuit, whereby different design choices or architectures may be considered. Hence, it is very useful to extract as many data as possible for parameter variability, especially related to different transistor geometries. A significant different parameter standard deviation has been observed for varying transistor geometries, resulting in the best variance for the largest TFT due to averaging of process variability on larger areas. This is a similar effect to that described by Pelgrom et al. for Si-CMOS technologies [176]. In this work, we have taken care of this effect by using two different TFT models for two transistor sizes during simulations, whereby each model has its own absolute value of parameter variability. Moreover, it is important to gather more data on TFT parameter variation at varying geometric sizes in order to include area dependent variability more accurately in simulations. The design of a flexible RFID transponder chip has been used as a case for WID variations. The final circuit yield at different supply voltages can be estimated well by MC simulations including area dependent variability and matches to the measurement results.

D2D variations on the other hand can be taken into account by performing corner analysis. Process corners are defined as slow, typical, or fast parameters for the transistor and are in this work defined as 3σ from the minimum and maximum average parameter among all dies considered. Corner simulations can be used to investigate whether the functionality of the circuit can be guaranteed or whether all specifications are met for all corners, power consumption, speed limitations, and so on.

Thin-film transistor technologies have the possibility for an individual back-gate contact for each transistor. This paves the way to implement adaptive back-gate control in the final circuit, resulting in increased robustness and therefore enabling a larger integration density and yield, hence reduced cost. Besides correcting for WID variations by applying a limited number of back-gate biases for the full circuit, adaptive back-gate control can also act as an excellent tool to shift the threshold voltages from corners toward typical values (or fast values) to accommodate for speed specifications. An option toward an integrated solution might be the implementation of a critical path replica tuning circuit.

In the final section of this chapter, variability issues specific to the field of large area electronics using the case of flexible AMOLED displays have been discussed. WID simulations will not be sufficient because of the size of the *circuit* (pixel engine) and back plane. On the pixel level, variations are taken into account by more complex pixel engines accommodating for variations arising before and during operation. The complexity of pixel engines is limited by the resolution of the display and the performance of the technology. A gate driver circuit enables the possibility of flexible displays. The design of this circuit will be influenced by the D2D variations, since the layout is spread out over many dies, eventually over the full wafer. Here, adaptive back-gate control could also be a beneficial option to comply with power, speed, and robustness specifications.

5 Design Case: RFID Tags

In this chapter we will elaborate on the specific design case of flexible, low-cost RFID tags. Many research activities in the field of organic electronics are focused on plastic RFID tags because of the potential of this technology for low-cost, large volume manufacturing. The goal of this design case is to demonstrate that despite the low intrinsic performance of organic electronics, plastic RFID tags could meet specifications and requirements set for Si-based RFID tags.

All the designs have been realized by following a full-custom, transistor-level design flow, taking into account the variability of the devices. In this design flow, mismatch simulations are performed on basic gates in order to obtain the most appropriate sizing while maximizing the robustness. Next, these basic gates are used in circuit building blocks, which are also evaluated using mismatch simulations. The final step is to perform such simulations for the targeted design. For more information, we would like to refer to the previous chapter, which elaborates on an example of mismatch simulations aiming for an 8-bit RFID transponder chip. The main logic family used for the designs in this chapter is unipolar, single V_T, zero-V_{GS}-load logic, determined by the depletion-mode characteristics of the transistors, so-called normally-on devices.

In 2007, Cantatore et al. [107] published a capacitive-coupled RFID system where a 64-bit code was read out at a base carrier frequency of 125 kHz. The 64-bit code generator was fully functional at a 30 V supply voltage. In that work, bit generators (up to 6 bit) could be read out using a base carrier frequency of 13.56 MHz by a capacitive antenna. The ability to use organic electronics for inductively coupled systems at 13.56 MHz has been shown by Böhm et al. [187], who demonstrated the readout of a ring oscillator with a clock frequency of 120 Hz using an inductive antenna. Ullmann et al. [108] demonstrated a 64-bit tag working at a bit rate exceeding 100 b/s at a base carrier frequency of 13.56 MHz. PolyIC has developed a 4-bit organic complementary RFID tag comprising n-type and p-type organic field-effect transistors at 13.56 MHz [117]. Jung et al. demonstrated a fully printed, roll-to-roll printable 1-bit RFID tag operational at 13.56 MHz [188]. More recently, the same group has increased the complexity of the transponder chip at the level of 16 bits [189]. The semiconductor for the latter printed tags is carbon nanotubes.

In this work, we will first discuss the road map for low-cost RFID tags and the Si-CMOS based standards. Afterward, we will demonstrate an inductively coupled, 64-bit RFID tag at a base carrier frequency of 13.56 MHz [97]. As a next step, the complexity of the transponder chip will be increased toward 128 bits of code, with the addition of a basic anti-collision protocol and write-once read-many times (WORM) memory [97]. In order to comply with standards set for Si-CMOS technologies regarding data rates, we investigate downscaling of the technology as a potential route to reach the specifications [118]. Alternative solutions are the integration of metal-oxide thin-film transistor technologies or the implementation of other logic families. Tripathi et al. have published the first 8-bit RFID transponder chip based on metal-oxide thin-film transistor technologies processed below 150 °C [99]. A new route to bi-directional communication for thin-film RFID tags will be discussed subsequently [101]. A bi-directional communication protocol needs to be established in order to read multiple tags present in the RF field. Before summarizing, we will detail an option to increase the robustness of thin-film RFID transponder chips by the addition of a back gate to each organic thin-film transistor in the circuit [115].

5.1 RFID and the Road Map for Low-Cost RFID Tags

RFID is an important technology in logistics, retail, automation, anti-counterfeit protection, identification, and several other fields. It allows for transmission of data (an identification code in the memory of the transponder chip) via radio waves from a transponder to a reader. In standard, silicon based electronics, this is a large market, segmented into several areas of applications. Several radio communication frequencies are used, spanning from tens of kilohertz to the low gigahertz range.

The goal of organic RFID tags is clearly the market of low-cost tags and labels. Therefore, we consider here only passive tags, which merely comprise an antenna and a transponder chip and need no battery (see Figure 5.1). The antenna of the tag is designed to resonate at the frequency of the electromagnetic wave emitted by the reader antenna. The transponder chip contains the following three functions:

1. A rectifier on the transponder chip produces a DC voltage from the electromagnetic wave captured by the antenna, with which the rest of the chip is powered up;
2. The bit sequencer reads out data from a non-volatile memory as a sequence of bits;
3. A modulator varies the impedance of the antenna synchronously with the bit sequence.

The reader can decode the data by sensing the modulation of the absorbed power of the reader antenna [190].

Item-level tagging – in which RFID replaces identification means such as bar codes on each retail item – is potentially a very large market for low-cost passive organic RFID tags. In that respect, a development goal for organic RFID technology is to become compatible with standards that have already been defined for that

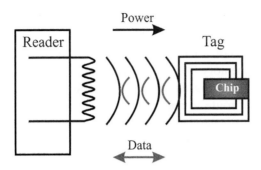

Figure 5.1   Schematic overview of a passive RFID system.

application. Indeed, this would ensure compatibility with installed infrastructure and future interoperability.

EPC Global is an organization entrusted by industry to create and support the Electronic Product Code (EPC). This is a global convention for immediate, automatic, and accurate identification of any single item in the supply chain of any company, in any industry, anywhere in the world. These conventions rely on the ISO norms, especially the ISO 18000 family. EPC is widely used already today, for example, on pallet level logistics. The next step is to use EPC tags at the package level, and the longer term target will be its use for individual items or, in other words, for item-level tagging.

One of the major limits of the EPC tags in their development to item-level tagging is the price of the single silicon transponder chip on the tag. The costs of the tags (or "inlays," according to a widespread naming system) are dominated by the cost of the silicon chip, followed by the cost of the antenna and of the assembly. For low-cost products, the cost of the silicon transponder chip is significantly too high. Lowering this cost is very challenging, because it is proportional to the chip size, which cannot be decreased at will because of the minimum size needed for economic handling and die attachment on the inlay.

Printed thin-film electronics could create a breakthrough in this respect. Lower cost per chip could result from the use of fewer process steps, less energy in the production (resulting from the lower process temperature), use of low-cost substrates (which saves packaging costs), and higher processing speeds, as well as larger area of base substrates (300 mm Si wafers versus Gen-6 display glass, for instance), resulting in a lower depreciation cost of manufacturing equipment per chip.

Furthermore, the availability of a thin-film circuit directly on a plastic substrate confers superior mechanical properties to the assembled tag. Conventionally, the lamination of the RFID "inlay" in cardboard, paper, or thin plastic results in an end-product with uneven topology, because of inlay thickness variations caused by the presence of the rigid silicon chip – and this limits the possibilities of roll-to-roll handling and lamination to products. Thinned-down silicon chips exist, but they are more expensive and often more fragile, limiting the throughput of die attachment into inlays. In contrast, thin-film chips are thin and flexible, qualities that solve both the handling and the yield issues.

Table 5.1. Main technology-related specs of EPC Gen 1 and Gen 2

Summary of EPC specs	EPC ISM band class 1	EPC class 1 generation 2
Carrier frequency (f_c)	13.56 MHz	860 MHz–960 MHz
Baud rate interrogator to label	26.48 kbps (f_c /512)	26.7 kbit/s–128 kbit/s
Baud rate label to interrogator	52.969 kbit/s (f_c /256)	Baseband modulated: 40 kbps– 640 kbps Subcarrier modulated: 5 kbps–320 kbps
Number of bits	64 or 96 + 16 bit redundancy + 24 bit destroy ("kill") code	Same, plus access password of 16 bit
Energy	Passive (no battery)	Passive (no battery)
Anti-collision	Label must be selectable in a group of related labels (interrogator talks first)	Label must be selectable in a group of related labels (interrogator talks first)
Destroy	Individually destroyable	Individually destroyable
Max. power of interrogator	Europe: 42 dB μA/m at 10 m	Europe: 42 dB μA/m at 10 m
Memory	Programmable	Programmable
Corresponding ISO/IEC norm	ISO/IEC 18000–3	ISO/IEC 18000–1, ISO/IEC 18000–6

Since item-level tagging may be the goal of organic thin-film RFID tags, and since EPC Global proposes standards applicable to the item-level tagging application, it is important to verify that the EPC standards can be reached with organic thin-film technology. Within EPC, several classes and generations can be distinguished. In Table 5.1, the main technology-related specifications of EPC systems at 13.56 MHz (HF) and 860-960 MHz (UHF) are listed for passive tags. The high-frequency (HF) and ultra-high frequency (UHF) classes are two commonly used bands for these RFID applications in the electromagnetic spectrum. HF is defined to cover all frequencies between 3 MHz and 30 MHz, while UHF represents frequencies between 300 MHz and 3 GHz.

At UHF communication frequency, the antennas can be printed, can make use of less-conductive metal tracks, and are smaller compared to that at HF. That implies that UHF allows for lower-cost antennas and is therefore the best solution for the lowest cost tags. However, the rectifier at UHF requires diodes with multi-gigahertz cut-off frequency, which have not been shown in thin-film technologies (organic or based on other semiconductors) so far.

Therefore, demonstrations so far concern HF tags. Their antenna is a resonant circuit comprising a coil with very low resistance and a capacitor. Several demonstrations of code generators, transponder chips, and HF tags have been shown and will be discussed in the next paragraphs.

A further point of consideration is the nature of the non-volatile memory containing the code. Implementations so far were hard-wired memories, generated at

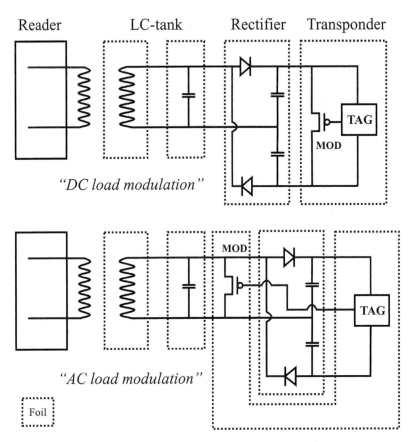

Figure 5.2   Inductively coupled organic RFID tags using DC (top) and AC (bottom) load modulation.

the site of production of the tag. Clearly, WORM (write-once read-many) memory could already be a great improvement on this, since the programming could be done at the site of use. A plastic WORM technology that could be envisaged in this respect is shown in [191] using fuses of poly(3,4-ethylenedioxythiophene) poly(styrenesulfonate) (PEDOT:PSS). Ultimately, an electrically re-programmable plastic memory compatible with and embeddable in thin-film plastic circuitry would be the most versatile solution. The required number of bits is modest, from 64 for the simplest tags, to kilobits for more elaborate versions. Several possible memory elements that could serve this purpose, including ferroelectric diodes and transistors, have been demonstrated already.

Finally, it should be noted that thin-film technologies are also ideal to make (large-area) sensors or actuators that can be integrated with RF tags. Furthermore, display effects such as electrophoretic, electrochromic, or even OLED can also be integrated with thin-film and organic transistors. The thin-film circuit technology can thus act as a versatile platform for smart labels that sense, display, and transmit information.

5.2   A Fully Integrated, 64-Bit Organic RFID Tag

The basic schematic of the organic RFID tag described in this subsection is depicted in Figure 5.2. The organic RFID tag consists of four different modules: the antenna coil, the HF-capacitor, the rectifier, and the transponder chip with an integrated load modulator. The coil and the HF-capacitor form an LC tank resonating at the HF resonance frequency of 13.56 MHz, which provides the energy for the organic rectifier with an AC voltage at 13.56 MHz. The rectifier generates the DC supply voltage for the 64-bit organic transponder chip, which drives the modulation transistor between the on- and off-states with a 64-bit code sequence. Load modulation can be obtained in two different modes, depending on the position of the load modulation transistor in the RFID circuit, shown in Figure 5.2. AC load modulation, whereby the modulation transistor is placed in front of the rectifier, sets demanding requirements to the OTFT, since it has to be able to operate at HF frequency. This is not obvious, as a consequence of the limited charge carrier mobility of the OTFT, which is 0.1 to 1 cm^2/Vs for pentacene as the organic semiconductor. Therefore, load modulation at the output of the rectifier (DC load modulation) is preferred in organic RFID tags. In the latter mode, the OTFT does not require operation at HF frequency. The organic RFID tags in this subsection operate in DC load modulation mode. Nevertheless, organic RFID tags operating in AC load modulation mode have also been achieved [97].

5.2.1   Technology

In this section we describe the technology used to create high-performance organic RFID tags [97]. As mentioned earlier, the tags are composed of four flexible foils, with the following components: an inductor coil, a capacitor, a rectifier, and a transponder. The coil is made from etched copper on foil, manufactured by Hueck Folien GmbH.

The HF-capacitor consists of a metal-insulator-metal (MIM) stack, processed on a 200-μm-thick flexible polyethylene naphthalate (PEN) foil (Teonex Q65A, Dupont Teijin Films). The insulator material used for the capacitor is Parylene diX SR.

The rectifier comprises two vertical Schottky diodes, and two capacitors in a so-called double half-wave configuration [192]. The schematic of the rectifier is shown in Figure 5.9, and a photograph of the rectifier is depicted in Figure 5.3. The substrate for manufacturing the rectifiers is a 200-μm-thick, flexible 150-mm-diameter PEN foil, on which first a metal-insulator-metal (MIM) stack is processed for the capacitors in the circuit. The metal layers are 30 nm of gold (Au) and the insulator is Parylene diX SR, with a relative dielectric constant of ε_r of 3 and a thickness of 400 nm. Conventional photolithography is used to define the capacitors in the MIM stack. The top Au layer of the MIM stack is used as the anode for the vertical diode. A 350 nm pentacene layer, the organic semiconductor, is evaporated through a shadow mask by HV-deposition. Finally, an aluminum (Al) cathode is evaporated through a second shadow mask.

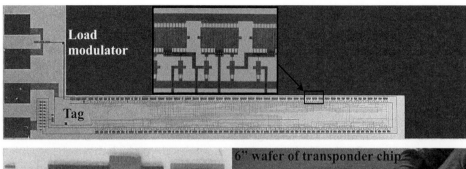

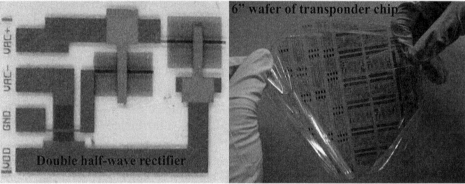

Figure 5.3   Pictures of the transponder foil and the load modulator foil (top), the double half-wave rectifier foil (bottom left), and the 150 mm flexible wafer full of transponder chips (bottom right).

The organic 64-bit transponder chip is made on a 25 μm thin plastic substrate using organic bottom-gate thin-film transistors. The organic electronics technology that is used was developed by Polymer Vision for commercialization in rollable active matrix displays and is described in Chapter 2, Section 2.4.1 [95]. A micrograph picture of the 64-bit transponder chip and the 150 mm wafer is also depicted in Figure 5.3.

5.2.2   RFID Measurement Setup

The complete tag is realized by properly interconnecting the contacts of the four foils, which we have achieved in an experimental setup where we plug the individual foils into sockets, as shown in Figure 5.4. Alternatively, we have also achieved tags by lamination of the foils, whereby electrically conductive glue is used to interconnect the different contacts of the individual foils.

The reader setup conforms to the ECMA-356 RF Interface Test Methods Standard. It comprises a field generating antenna and two parallel sense coils (Figure 5.4), which are matched to cancel the emitted field. By this method, only the signal sent by the RFID tag is read out at the reader side. The detected signal is then demodulated by a simple envelope detector (inset Figure 5.11), which is a diode followed by a capacitor and a resistor, and shown on an oscilloscope.

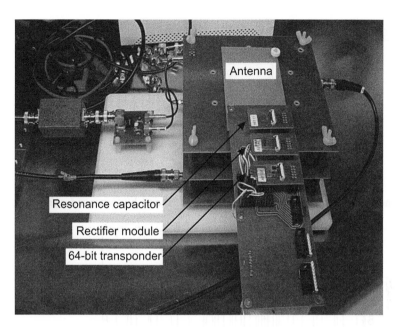

Figure 5.4   Overview of the reader and RFID tag measurement setup. Foils are placed in sockets to ease manipulation.

5.2.3   Organic Transponder Chip

Possible logic types in unipolar, single threshold voltage logic are zero-V_{GS}-load logic and diode-load logic, as explained in Chapter 3. The gates in our organic RFID transponder chip are designed using the former, which is also known as zero-V_{GS}-load logic. The choice for this type of logic was driven by the fact that the transistors used in this work show normally-on or depletion-mode behavior. The final design of the organic transponder chip has been made using only inverters and NAND-gates, both implemented in the zero-V_{GS}-load logic. The gains of such an inverter at the trip point, for supply voltages of 10 V and 20 V, are 1.75 and 2.25, respectively. Nineteen-stage ring oscillators of inverters operate at a frequency of 627 Hz when using 10 V supply voltage and at 692 Hz when using 20 V supply voltage. Both the channel length and the FW (finger width) are 5 μm.

The schematic of the transponder chip is depicted in Figure 5.5. A 19-stage ring oscillator generates the clock signal when powered. This clock signal is used to clock the output register, the 3-bit binary counter, and the 8-bit line select. The 8-bit line select has an internal 3-bit binary counter and a 3-to-8 decoder. This block selects a row of 8 bits in the code. The 3-bit binary counter drives the 8:1 multiplexer, selecting a column of 8 bits in the code matrix. The data bit at the crossing of the active row and column is transported via an 8:1 multiplexer to the output register, which sends this bit on the rising edge of the clock to the modulation transistor. In this design, the code is hardwired. The 3 bits of the 3-bit binary

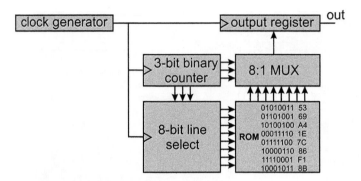

Figure 5.5   Schematic overview of the digital logic part of the 64-bit transponder chip.

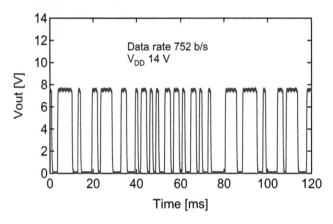

Figure 5.6   Measured code of the 64-bit transponder chip at a supply voltage of 14 V.

counter are also used in the 8-bit line select block for selecting a new row after all 8 bits in a row are transmitted.

The 64-bit transponder foil comprises only 414 organic thin-film transistors. Figure 5.3 shows a micrograph image of the transponder foil. At 14 V supply voltage the 64-bit transponder foil generates the correct code at a data rate of 752 b/s, depicted in Figure 5.6. Besides the 64-bit transponder foil, we designed an 8-bit, 16-bit, and 32-bit transponder foil. The main difference in the designs is the complexity of the line select. The measured data rates of these transponder foils are plotted in Figure 5.7. The obtained measurement results correspond to the yield prediction discussed in Chapter 4 for this design. These transponders are operational at 14 V supply voltage, with a minimum supply voltage for some circuits of 10 V.

5.2.4   Organic Rectifier

The purpose of the rectifier in an RFID tag is to create a DC voltage from the AC voltage detected and generated by an antenna at the targeted base carrier frequency

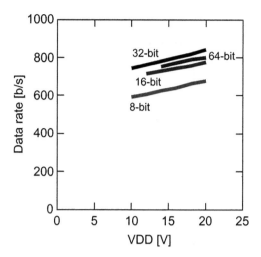

Figure 5.7   Data rates of measured 8-bit, 16-bit, 32-bit, and 64-bit RFID transponder chips plotted as a function of supply voltage.

of 13.56 MHz. This frequency is selected because it is a standard in Si-based RFID tags and will therefore enable partial compatibility with installed reader systems at 13.56 MHz. An important issue for organic RFID tags is the efficiency of the rectifier. A more efficient rectification generates the required DC voltage from a smaller AC input voltage. This results in larger reading distances for the RFID tags [97], [192].

A rectifier comprises diodes and capacitors. For organic diodes, two different topologies can be used: a vertical Schottky diode [193], [194] or a transistor with its gate shorted to its drain. The transistor with shorted gate-drain node is often considered the most favorable topology because its process flow is equal to that used for the transistors in the digital circuit of the RFID tag. We have chosen to use the vertical diode structure because of its better intrinsic performance at higher frequencies compared to transistors as diodes [195]. As an alternative option, metal-oxide diodes and rectifiers have already been demonstrated for Schottky [196] and transistor topologies [99], [197].

The structure of a vertical organic diode used to make the rectifier is shown in Figure 5.8 (a). As depicted, we fabricated hole-only organic diodes with a layer of pentacene (as organic semiconductor) sandwiched between an Au- and an Al-electrode on a 150 mm PEN foil. Because of their work functions, the Al-electrode blocks the injection of holes, whereas the Au-electrode permits the injection of holes. The current-voltage characteristics of such a pentacene diode are depicted in Figure 5.8 (b). The plotted characteristics show an onset voltage of the diode at about 1.2 V. At a 3 V forward bias, the current density is 2.88 A/cm^2 for a device with an active area of 500 μm × 200 μm.

We have fabricated a more efficient rectifier circuit, a double half-wave rectifier. This rectifier comprises two diodes, each followed by a capacitor [192]. Figure 5.9

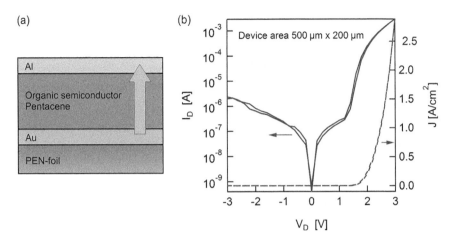

Figure 5.8   (a) Structure of a vertical, organic semiconductor-based Schottky diode; (b) |I|-V and J-V characteristics of an organic pentacene diode on logarithmic (left axis) and linear (right axis) scales.

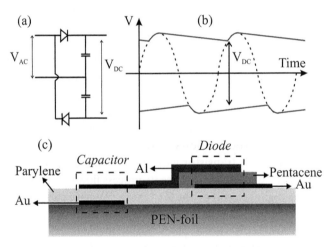

Figure 5.9   (a) Schematic of a double half-wave rectifier, (b) schematic operation of a double half-wave rectifier, (c) vertical cross section of the integrated capacitors and diodes on foil.

(a) shows the schematic of this circuit. Both capacitors are 20 pF. The active area of the diodes is 500 μm × 200 μm.

A double half-wave rectifier circuit consists of two single half-wave rectifiers connected between the same nodes, with diodes connected as shown in Figure 5.9 (a). Both single half-wave rectifiers rectify the AC input voltage: one rectifies the upper cycles of the AC input voltage; the other single half-wave rectifier rectifies the lower cycles of the input voltage. This is schematically depicted in Figure 5.9 (b). The power and the ground voltage for the digital logic of the RFID tag are taken between both rectified signals (Figure 5.9 (a) and (b)). Therefore, a double

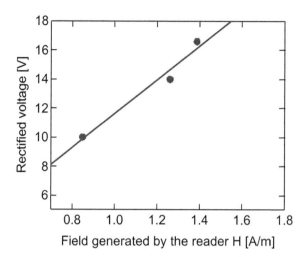

Figure 5.10 Internal rectified voltage of a double half-wave rectifier generated in an organic RFID tag versus the 13.56 MHz magnetic field generated by the reader.

half-wave rectifier yields about double the rectified voltage compared of a single half-wave rectifier.

5.2.5 Organic RFID Tag Using DC Load Modulation

The organic RFID transponder foil, the antenna coil, the HF-capacitor, and the rectifier on foil are connected to form an organic RFID tag. In DC load modulation mode, the modulation transistor (W/L = 5040/5) is placed behind the rectifier, as can be seen in Figure 5.2. All foils are placed into sockets and connected as depicted in Figure 5.4. The RFID reader is a 7.5 cm radius antenna, which emits the field at a base carrier frequency of 13.56 MHz. In Figure 5.10, the internal rectified voltage of this double half-wave rectifier in the organic RFID tag is plotted as a function of the field generated by the reader, for the tag antenna placed in the near-field of the reader antenna, at a distance of about 4 cm from the coil generating the reader's 13.56 MHz RF field. As can be seen in the graph, 10 V rectified voltage is obtained at a 13.56 MHz electromagnetic field of about 0.9 A/m, and 14 V at 1.26 A/m. The latter is the voltage required by the 64-bit organic transponder chips. The ISO 14443 standard states that RFID tags should be operational at a minimum required RF magnetic field strength of 1.5 A/m. The double half-wave rectifier circuit presented here therefore satisfies this ISO norm. After extrapolation of the measurement data, a DC voltage of 17.4 V can be obtained at a field of 1.5 A/m. If a single half-wave rectifier were used, the rectified voltage would be limited to 8–9 V, which is too low for current organic technology.

The rectified 14 V obtained drives the transponder chip, which sends the code to the modulation transistor. The signal sent from the fully integrated plastic tag is

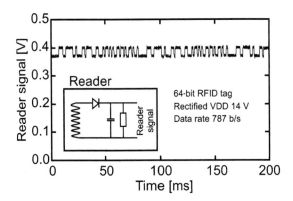

Figure 5.11   Signal of the 64-bit RFID tag measured on the reader (unamplified reader signal). The envelope detector of the reader is depicted in the inset.

received by the reader and subsequently visualized using a simple envelope detector (see inset Figure 5.11) without amplification. The signal measured at the reader side is depicted in Figure 5.11. This shows the fully functional 64-bit RFID tag using an inductively coupled 13.56 MHz RFID configuration with a data rate of 787 b/s. With a 0.7 V drop over the diode at the reader (envelope detector), a tag-generated signal of about 1.1 V is obtained, from which 30 mV is load modulation (modulation depth $h = 1.4$ percent).

Two of the reader standards at 13.56 MHz base carrier frequency are the proximity (ISO 14443) and vicinity readers (ISO 15693). The main difference between them is the coil radius, which is 7.5 cm for the proximity reader and 55 cm for the vicinity reader. This results in a maximum readout distance of 10 cm for the proximity and 1 m for the vicinity reader. As mentioned earlier, the standard (ISO 14443) states also that the tag should be operational at an RF magnetic field of 1.5 A/m, which is significantly lower than the maximum allowed RF magnetic field of 7.5 A/m. One can calculate the required magnetic field at the antenna center in order to obtain the required field to operate the tag. In our case, the required field for an 8-bit organic RFID tag was 0.97 A/m. This is depicted in Figure 5.12. The dots in this graph show the experimental data at distances of 3.75, 8.75, and 13.75 cm with respect to the field generating antenna. This graph shows that it is possible to energize the 8-bit organic RFID tag at maximum readout distances for proximity readers below the maximum allowed RF magnetic field. The signal detected by the reader during the same experiment is depicted in Figure 5.13 at distances of 5 and 10 cm with respect to the sense coil.

5.3   Adding More Complexity to the Transponder Chips

This section describes the continuation of the work toward EPC-compatible RFID tags whereby the complexity of the transponder chips needs to be increased. In this generation of transponder foils, 8- and 128-bit transponder chips having more functionality, including data encoding and a basic anti-collision protocol, were

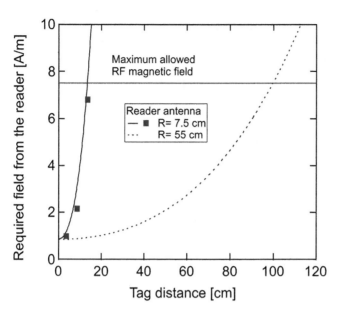

Figure 5.12 Calculation and experimentally obtained data of the required RF magnetic field at the reader side as a function of the tag distance in order to generate the required RF magnetic field to operate the tag.

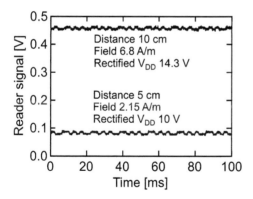

Figure 5.13 Signal of the 8-bit RFID tag measured on the reader (unamplified reader signal) in DC load modulation mode at distances of 5 and 10 cm.

designed [97] and fabricated using the technology described in Chapter 2, Section 2.4.1. This is the same technology used for the 64-bit transponder chip described in Section 5.2.1; however, the mobility and threshold voltage of the single transistors are slightly higher. A larger threshold voltage leads to a more normally-on transistor. This results in a 33-stage ring oscillator with a frequency of 1.8 kHz at 15 V supply voltage. The speed of the ring oscillator, and therefore also the bit rate of the transponder chip, is faster compared to generation 1 because of the higher mobility and threshold voltage of the resulting organic transistors.

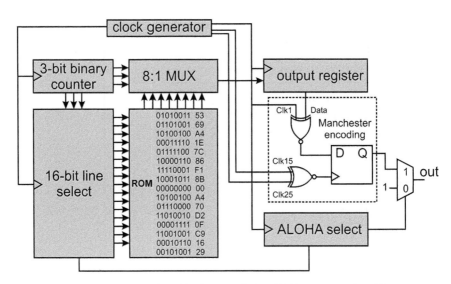

Figure 5.14   Schematic overview of the digital logic part of the 128-bit transponder chip.

The schematic overview of the 128-bit transponder chip can be seen in Figure 5.14. The difference between the 128-bit and 64-bit transponder chips (Section 5.2.3) is the complexity of the line select, which is now a 16-bit line select. Also here, 128 bits are hardcoded in the chip. The clock signal in the generation 2 transponder chips is a 33-stage ring oscillator, because the added functionality increases the critical path.

Data encoding is added to this second generation 8- and 128-bit transponder chip, more specifically Manchester encoding. A zoom in the digital logic of the Manchester encoding block is depicted in Figure 5.14. Manchester encoded data require, besides the normal bit transitions, a transition in the middle of the bit. A transition of 0 to 1 corresponds with a logic 0, and vice versa. In our design, every transition needs a rising edge of the clock. To include Manchester encoding in this scheme without losing data rate, a clock having double frequency is gener-ated by the 33-stage ring oscillator. This clock is used to encode the data. The gen-eration of this clock is done by an EXNOR behind stages 15 and 25, as depicted in Figure 5.14. The two measured clock signals are shown in Figure 5.15, whereby the speed of the Manchester clock (3.6 kHz) is double the speed of the system clock (1.8 kHz). The data are subsequently encoded by adding another EXNOR and NAND gate.

To enable the readout of multiple organic RFID tags at once, a basic anti-collision protocol is added to the plastic RFID transponder chip. The anti-collision protocol used is a basic version of ALOHA, which is a "tag talks first" protocol. A tag sends its code, after which a silent period follows. The code is then retransmitted. During the silent period, another tag can be read out. If a tag transmits its code during the transmitting time of another tag, a collision occurs and the code is consequently not valid. A full ALOHA protocol should also allow the reader to acknowledge the

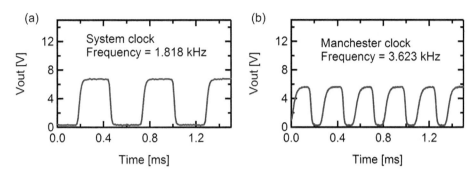

Figure 5.15   Measured signal of (a) the 33-stage ring oscillator, which is the system clock for these transponder chips, and (b) the generated Manchester clock. The supply voltage for these clocks is 15 V.

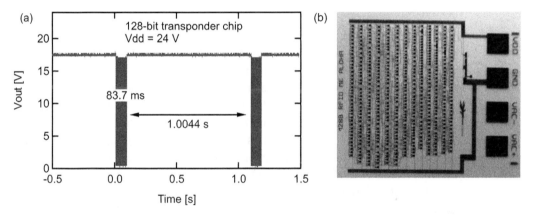

Figure 5.16   (a) Measured signal of the 128-bit organic transponder chip with a power supply of 24 V and (b) depiction of the 128-bit transponder foil whereby the functional area is about 1 cm².

successful detection of the code, after which the tag remains silent. This has not been implemented here because of the nonexistent communication from the reader to the tag.

In the implementation of this ALOHA protocol, a 1100 binary counter is used to select whether the data from the Manchester encoder should be sent out (value of the counter is 0000) or the supply voltage should be connected to the load modulator (all other values of the counter). The clock for this counter is generated by the 3-bit binary counter and the 16-bit line select. In this way, the silent period takes 12 times the time necessary to stream out all data bits. Figures 5.16 and 5.17 depict the measurement results of the 128-bit transponder chip, including Manchester encoding and the ALOHA protocol. This chip was powered with a supply voltage of 24 V and employs 1286 organic p-type transistors. The 128 bits can be read out in 83.7 ms, that is, a bit rate of 1529 Hz. Figure 5.16 also shows a picture of the 128-bit transponder foil.

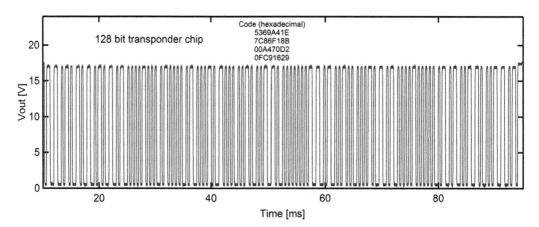

Figure 5.17   Zoom of 1 period of Figure 5.16, where the full code of the 128-bit organic transponder chip is shown.

Moreover, in the 8-bit RFID transponder chip, we exchanged 4 ROM bits with WORM memory. One bit of the latter memory is a zero-V_{GS}-inverter, whereby the pull-up transistor is connected to the ground and whereby the connection between the pull-up transistor and the VDD is a metal line, which can be interrupted by means of mechanical breakage (cutting) or thermal ablation (e.g., by a laser pulse), as can be seen in Figure 5.18. When this line is not interrupted, this inverter yields a logic 1, and after cutting this line, the bit changes to 0. Figure 5.18 shows the measurement results of the 8-bit transponder chip, including Manchester encoding and ALOHA protocol, when no bits are fused and when bit 5 is fused. This transponder chip is fully operational at a supply voltage of 27 V with a bit rate of 1583 Hz.

5.4   Can We Meet the Data Rate Targets for EPC Transponder Chips?

As described in Section 5.1, thin-film RFID tags need to be compatible with Electronic Product Coding (EPC), developed for high-volume logistics applications in order to find wide acceptance. In previous sections, we have shown that some of those EPC specifications have already been met by plastic tags in recent years, namely, transmission of 64-bit codes, as well as the use of HF (13.56 MHz) base carrier frequency compatible with regulations concerning human exposure to electromagnetic fields and basic anti-collision protocols. Nevertheless, state-of-the-art organic transponder chips have shown an order of magnitude lower data rate than required by the specifications (26.48 kb/s for the forward link and 52.969 kb/s for the return link [198]). The critical factor to address the circuit speed is the current drive of the transistors, which is determined by

- the carrier mobility,
- the specific capacitance of the gate dielectric,
- the inverse of the channel length.

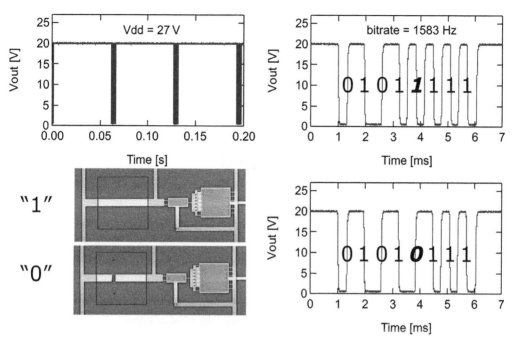

Figure 5.18   Measured signal of the 8-bit organic transponder chip including 4 bits of WORM memory, before and after switching of bit 5. This transponder chip is fully operational at a power supply of 27 V. A micrograph image of 1 bit WORM memory (in the initial state and after cutting) is also shown.

In this section, we will demonstrate several options to realize transponder chips fulfilling the latter EPC specification. The first option uses downscaling of channel length and parasitics for an organic RFID transponder chip; a second option integrates better-performing metal-oxide TFTs into a transponder chip. The last option discusses the implementation of other logic families compared to unipolar zero-V_{GS}-load logic, namely, diode-load logic and complementary logic.

5.4.1   Organic Transponder Chips

The technology for these RFID transponder chips is described in Chapter 2, Section 2.4.3. The transistors are normally-on and their charge carrier (hole) mobility exceeds 0.5 cm²/Vs. The oxide capacitance is larger for this technology (100 nm thick Al_2O_3) compared to that of chips fabricated in previously discussed technologies (Chapter 2, Section 2.4.1). Therefore, this technology allows downscaling of the transistor channel length, within the boundaries achievable by existing high-throughput tools (e.g., steppers used in back plane manufacturing). Specifically, we investigated downscaling of the channel lengths (L) of the circuits from 20 μm and 2 μm. This results in an output resistance in saturation at 0 V gate-source voltage of 207 ± 8 MΩ for $L = 5$ μm and 9.8 ± 0.2 MΩ for $L = 2$ μm.

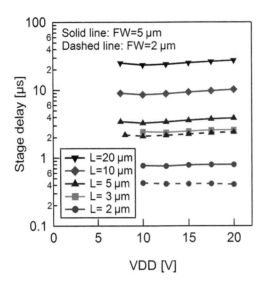

Figure 5.19   Overview of stage delays versus supply voltage, measured on 19-stage ring oscillators with varying channel lengths, between 20 μm and 2 μm (color and symbol codes) and source/drain finger sizes (*FW*) of 5 μm (solid lines) and 2 μm (dashed lines).

We also limited the parasitic gate-source and gate-drain overlap capacitances by decreasing the width of the finger-shaped source and drain contacts (which completely overlap the gate, since current technologies are not fabricated using self-aligned process steps, explained in Section 2.3 of Chapter 2) from 5 μm to 2 μm.

Figure 5.19 depicts the inverter stage delay as a function of supply voltage, as extracted from 19-stage ring oscillators, for transistors with channel lengths varying from 20 μm to 2 μm and with gate overlap of the transistor-fingers ranging from 5 μm to 2 μm. Stage delays below 1 μs, and as low as 400 ns, are shown at *VDD* as low as 10 V. The effect of decreasing the overlap capacitance is also shown in Figure 5.19 for the circuits having a channel length of 2 μm and 5 μm: shrinking the overlap from 5 μm (solid lines) to 2 μm (dashed lines) improves the stage delay by a factor of 1.5 to 2 [118].

We proceeded with the design and realization of 8-bit RFID transponder chips, having a channel length of 2 μm and either 5 μm or 2 μm finger widths. Figure 5.20 shows the photographs of the 150 mm foil-on-carrier wafer and a zoom of a single die. Figure 5.21 depicts the output signal of both types of transponders. In agreement with the two-fold faster inverter stage delay for the 2 μm fingers, the data rate of this transponder is also twice as high as that of the design with 5 μm fingers. The data rate obtained of the 8-bit transponder with channel length and fingers of 2 μm reaches 50 kb/s [118].

5.4.2   Metal-Oxide NFC Chips

Another possibility to obtain faster data rates, even without the necessity of downscaling, is the use of a semiconductor with intrinsically higher charge carrier

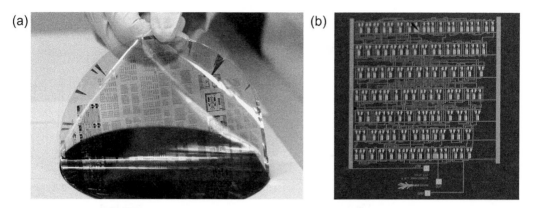

Figure 5.20   (Left) Photograph of the 150 mm foil on a carrier wafer comprising all measured circuits; (right) photograph of the 8-bit RFID transponder chip on foil. The die size is 24.73 mm². The design comprises 294 transistors.

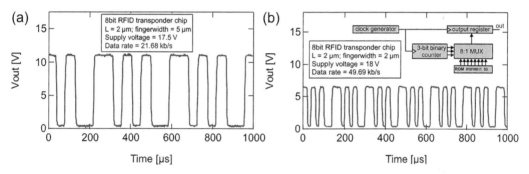

Figure 5.21   Measured signal of the 8-bit RFID transponder chip having transistors with channel length of 2 μm: (a) width of source-drain fingers of all transistors is 5μm; (b) width of source-drain fingers of all transistors is 2 μm. The supply voltages for the 8-bit transponder chips were (a) 17.5 V and (b) 18 V, and data rates are, respectively, 21.68 kb/s and 49.69 kb/s. The inset of the bottom picture shows the schematic overview of the digital logic portion of the 8-bit transponder chip.

mobility. Indium-gallium-zinc-oxide is a very promising n-type semiconductor that can be processed at low temperatures enabling a-IGZO transistor technologies on foil. This technology is described in Chapter 2, Section 2.4.4, and also features the high-k Al_2O_3 gate dielectric. The resulting transistors have mobilities around 9 cm²/Vs and excellent WID variation characteristics, as discussed in 2.4.4. Figure 5.22 (b) plots five zero-V_{GS}-load inverters measured at 10 V.

The design of the 8-bit transponder chip has been adapted for n-type semiconductors. The 8-bit RFID transponder chip based on a-IGZO zero-V_{GS}-load logic yields a data rate of 34.7 kb/s at a supply voltage of 10 V, having a channel length and source-drain finger width of 5 μm. In order to obtain data rates fully compatible with EPC regulations (52.969 kb/s), the technology can still be improved with slightly higher mobilities, or it can be realized with modest downscaling.

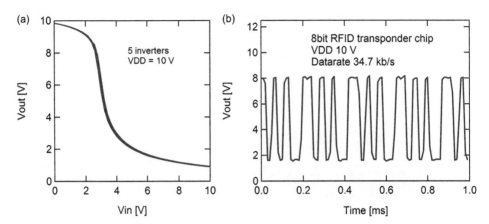

Figure 5.22   (a) Five zero-V_{GS}-load inverters at a supply voltage of 10 V and (b) measured signal of the 8-bit RFID transponder chip having transistors with channel length of 5 μm and finger width of source-drain fingers of 5 μm. The supply voltage was 10 V yielding a data rate of 34.7 kb/s.

5.4.3   Faster Transponder Chips by Other Logic Types

The realization of circuits by implementing other logic families provides another valid possibility for increasing the circuit's speed as necessary to comply with specifications. Figure 5.23 depicts the direct comparison of inverter stage delays when implemented in zero-V_{GS}-load logic, diode-load logic, and complementary logic. These measurements have been obtained on the same wafer. The ring oscillators are designed in a hybrid organic/metal-oxide complementary technology described in Chapter 2, Section 2.4.5. The unipolar ring oscillators are integrated using the n-TFTs in this technology. Figure 5.23 demonstrates experimentally that complementary and diode-load logic yield similar inverter stage delays, both faster with respect to zero-V_{GS}-load inverters.

The robustness for diode-load logic may be the limiting factor for implementation in larger circuitry as concluded from Chapter 3. Dual-gate (dual-V_T) diode-load logic exhibited improved robustness of this logic type enabling the integration into complex circuits. We show examples of transponder chips using diode-load logic in Section 5.6, realized in dual-gate pentacene technology described in Chapter 2, Section 2.4.2. Implemented with IGZO as semiconductor, these would also have resulted in circuits operating above 50 kb/s.

Complementary logic profits both from robustness and speed perspective. This provides as a consequence the best option to realize transponder chips matching the specifications. In Section 5.5 of this chapter, we have integrated transponder chips in this hybrid organic/complementary technology, operating already at 3.75 V supply voltage. This chip yielded data rates of 20 kb/s at 10 V supply voltage.

We have also investigated the impact of transistor channel length scaling and reduction of the parasitic capacitances on the circuit's performance. The technology for this study is based on a hybrid complementary technology, but using an

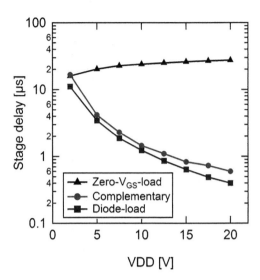

Figure 5.23 Comparison of inverter stage delays when implemented in 19-stage ring oscillators for different logic families, n-TFT zero-V_{GS}-load, n-TFT diode-load, and complementary logic, measured on the same wafer for a channel length of 5 μm.

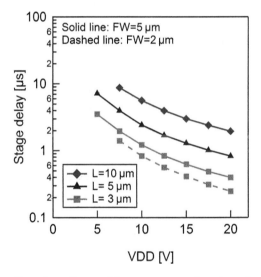

Figure 5.24 Overview of stage delays versus supply voltage in a complementary technology, measured on 19-stage ring oscillators with varying channel lengths, between 10 μm and 3 μm (color and symbol codes) and source/drain finger sizes (*FW*) of 5 μm (solid lines) and 2 μm (dashed lines).

alternative metal-oxide semiconductor enabling integration on PEN-foil. This technology is described in Chapter 2, Section 2.4.6. Figure 5.24 plots the inverter stage delays for the latter technology based on measurements of 19-stage ring oscillators. The channel lengths have been decreased from 10 μm to 3 μm, improving the stage delay between 5.63 μs and 1.21 μs, respectively, at 10 V supply voltage. The impact of reducing the parasitic overlap capacitance has been demonstrated by decreasing the source-drain finger width (FW) from 5 μm to 2 μm. This yields a 1.46-fold improvement in stage delay.

5.5   Bi-Directional Communication

A key specification of EPC is the demand of bi-directional communication to avoid collision of data at the reader when multiple tags are present simultaneously in the field of the reader. The protocol of data transfer from tag to reader specifies that the reader should talk first and give instruction to the tags one by one to transmit their code. This requires the necessity to demodulate, at the tag side, information transmitted by the reader. Furthermore, the data clock should be derived from the HF field, and not be generated as an independent and asynchronous clock on the tag. All thin-film RFID tags reported prior to 2012, however, were based on a tag-talks-first principle: as soon as the tag is powered from the RF field, its code is transmitted at a data rate determined by an internal ring oscillator. Practical RFID systems will need to be able to read multiple RFID tags within the reach of the reader antenna. Existing anti-collision protocols implemented in organic RFID tags [97] are limited to about maximum 4 tags at the expense of a slow reading time. In this section, we demonstrate a reader-talks-first low-temperature thin-film transistor RFID circuit [101], [201].

The designs have been realized in a complementary hybrid/metal-oxide technology on rigid substrates, described in Chapter 2, Section 2.4.5. The p-TFT and n-TFT currents are matched at an optimal 2:1 ratio. A typical VTC of the inverter is shown in Figure 5.25. The inverter is operational from 2.5 V onward. As organic transistors with reasonable channel lengths (≥ 2 μm) have a cut-off frequency below 13.56 MHz, the base carrier frequency of HF communication, present technologies on foil do not yet allow extraction of the circuit clock as a fraction of the base carrier. We solve this by introducing an original uplink (reader-to-tag) scheme, in which a slow clock (compatible with our transistor's speed) is transmitted as amplitude-modulation on the base carrier while data are encoded on this clock by pulse width modulation (PWM).

Low-cost tags will need to be passive, that is., draw the power for the circuit from the RF field by means of a rectifier operating at HF (13.56 MHz). We designed double half-wave transistor rectifiers [192], [195] using solution-processed metal-oxide TFTs. These rectifiers operated up to frequencies beyond 100 MHz, owing to both the high mobility (μ > 2 cm^2/Vs) and the low zero-V_{GS} current of the oxide TFTs (see Figure 5.26). Two different double half-wave transistor rectifiers are used in our tag design (see Figure 5.27). A slower rectifier provides the DC power voltage to the tag. The faster rectifier is designed with a time constant faster than the amplitude-modulated clock sent in the uplink, and thus allows detection of this uplink data clock as explained further.

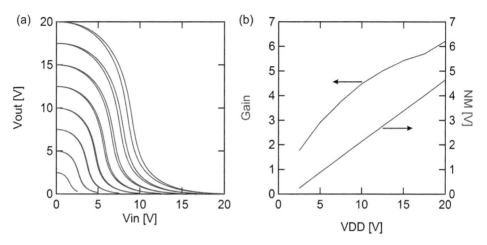

Figure 5.25   (a) Typical VTC of a hybrid organic/metal-oxide inverter with a p:n ratio of 2:1, measured at different supply voltages ranging from 2.5 V to 20 V in steps of 2.5 V; (b) the extracted gain and noise margin (NM) are plotted as a function of the supply voltage.

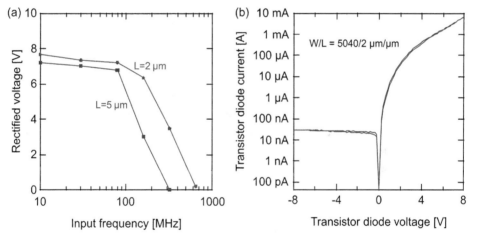

Figure 5.26   (a) Rectified voltage of double half-wave rectifiers as a function of frequency at a 5 V AC amplitude. The rectifiers comprise two capacitors and two solution-processed oxide transistors. (b) The current voltage characteristics of the oxide transistor-diode.

In order to obtain bi-directional data communication with the uplink protocol discussed, a dedicated reader has been developed that allows amplitude modulation for the up-link data transfer using pulse width modulation, that is, having a different duty cycle for logical 0 (75 percent active low) and 1 (25 percent active low) in the different code sequences.

The overall schematic of the tag is shown in Figure 5.27. The tag comprises a resonant antenna at 13.56 MHz, two different rectifiers, a 4-bit input decoder circuit that generates a hit/kill signal (comparator), an 8-bit code sequencer, and a load modulator. When the hit signal is active, the 8-bit code sequencer starts

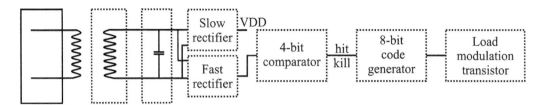

Figure 5.27   Schematic overview of the different building blocks of the hybrid RFID tag.

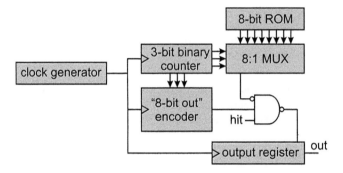

Figure 5.28   Block diagram of the 8-bit code generator.

generating 8 data bits, followed by a sequence of 8 digital ones. Figure 5.28 depicts the corresponding block diagram of the 8-bit code sequencer. Two different code generators have been designed and measured with two different codes (code 1 = 01010011, code 2 = 010111001). Figure 5.29 shows the correct behavior of the code generator when the hit signal turns on and off. The data rate is also plotted as a function of the supply voltage. The code generator is operational at a supply voltage as low as 3.75 V, resulting in a maximum data rate of 20.6 kbit/s at a VDD of 10 V. The die picture of this code sequencer is shown in Figure 5.29 (c); its size equals 5983 × 6064 μm².

Figure 5.30 shows the details of the data extractor of the input decoder. It extracts both the data and the clock from the rectified RF signal. After three inverter stages for restoring the input signal to the logic levels, a dedicated delay gate is used (Δt in Figure 5.30). This delay gate consists of a pull-up transistor with W/L of 40/10 and a 50/50 pull-down transistor. The logic 1 (25 percent active high on after three inverter stages) will discharge the 106 pF capacitor much less compared to the logic 0 (75 percent active high on after three inverter stages) and hence a different signal is stored in the D-Flip-Flop at the next falling edge of the incoming signal. The correct behavior of the delay inverter is analyzed using the delay inverter test circuit, as depicted in Figure 5.31. The measurements obtained visualize a larger delay when the input switches from low to high compared to a high-to-low switch.

Two different 4-bit input decoders (code A = 0010, code B = 0110) have been designed and tested. Figure 5.32 (b) depicts the output signal of both comparators

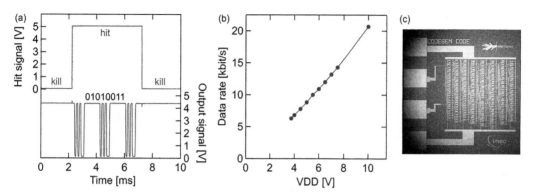

Figure 5.29   (a) The output signal of code generator 1 (01010011) when the hit is active at a supply voltage of 5 V. (b) The data rate of this code generator plotted as a function of the supply voltage and (c) die photograph of this code sequencer.

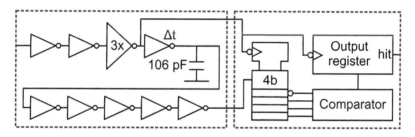

Figure 5.30   Detailed schematics of the comparator: the uplink data extractor and shift register.

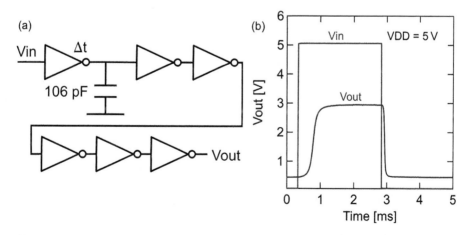

Figure 5.31   (a) Schematics of the delay inverter test structure and (b) its corresponding measurement results at a supply voltage of 5 V.

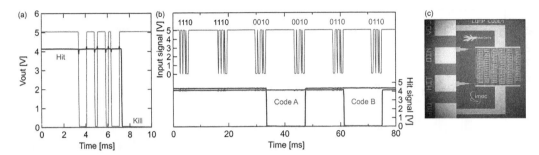

Figure 5.32 (a) A zoom of the output signal of the comparator toggling the hit signal at a supply voltage of 5 V. (b) The output signal of both comparators demonstrating the selectivity to the input stream and (c) die photograph of a comparator.

to an input stream with different codes. This plots the selectivity of both comparators. Figure 5.32 (a) shows a zoom of one of the comparators toggling the hit signal after verification of the correct input code. The die picture of this code sequencer is shown in Figure 5.32 (c); its size equals $3579 \times 4228 \ \mu m^2$.

Figure 5.33 (a) shows the internal RFID tag voltages during system testing using a setup as indicated in Figure 5.27. An input decoder for detecting the uplink code A and an output sequencer for generating code 2 are used during this test. The external RFID reader can toggle the code generator on Figure 5.33 (b) and off (c). One observes that the internal power voltage of the tag (slow rectifier) substantially drops during amplitude modulation at the reader, but the use of hybrid CMOS still ensures the correct operation at these low supply voltages. The overall bi-directional tag comprises 368 metal-oxide transistors and 365 pentacene transistors closely integrated on the same substrate.

To summarize this section: we have demonstrated a viable route toward bi-directional communication at 13.56 MHz by low-cost RFID tags in a complementary, hybrid solution-processed metal-oxide/organic thin-film transistor technology. As these transistors do not allow one to decode a clock directly from the HF base carrier, the up-link clock is transmitted as amplitude modulation of the carrier, and uplink data by pulse width modulation of this clock. Our solution will enable true anti-collision protocols for low-cost HF RFID tags. Moreover, fast data rates for the code generator are shown using this technology, making such hybrid complementary technologies viable candidates for RFID tags at EPC data rates (see Section 5.4).

5.6   RFID Transponder Chips with Increased Robustness

As we discussed in much detail in former chapters, dual-gate thin-film transistor technologies can be used to increase the robustness of logic gates, as a result of an increased noise margin [115]. In Chapter 3, we have detailed an optimized dual-gate inverter topology for both zero-V_{GS}-load and diode-load configurations based on

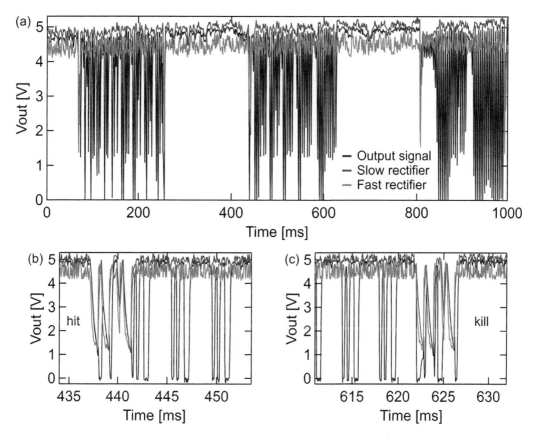

Figure 5.33   (a) Measurements of the internal voltages of the bi-directional hybrid RFID tag when powered by the 13.56 MHz signal of the reader; (b–c) a zoom of the hit and the kill transients.

the technology described in Chapter 2, Section 2.4.2. The advantages of the optimized configurations are increased noise margins for both topologies and the necessity of only one extra voltage rail, namely, the bias voltage to the back gate of the drive transistor. In this section, we integrated the optimized configurations into ring oscillators and RFID transponder chips [115]. We also would like to refer to the next chapter, where the same technology has been used to fabricate the first 8-bit, flexible organic microprocessor [63].

In order to demonstrate that organic p-type only dual-gate technology can obtain a high integration density, we made 99-stage dual-gate ring oscillators with zero-V_{GS}-load and diode-load architectures. In Figure 5.34, we show the extracted stage delays of NAND gates and inverters, using the optimized dual-gate zero-V_{GS}-load and diode-load topologies as discussed in Chapter 3, and compare them to reference single-gate zero-V_{GS}-load gates fabricated on the same foil [115]. The fastest family is the optimized diode-load topology, with a stage delay of 2.27 µs. The optimized zero-V_{GS}-load topology is an order of magnitude slower (26 µs),

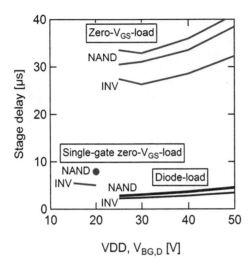

Figure 5.34   The stage delay of 99-stage ring oscillators is plotted as a function of the V_T-control voltage of the drive transistor ($V_{BG,D}$) for optimized zero-V_{GS}-load and diode-load inverters at a supply voltage of 20 V. The stage delay of a single-gate zero-V_{GS}-load inverter is shown for reference. This stage delay is plotted as a function of the power supply voltage (VDD).

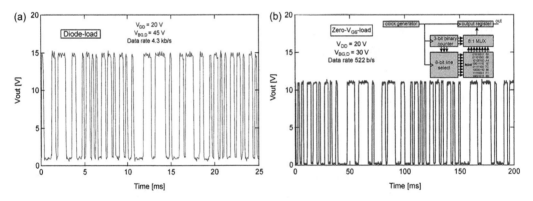

Figure 5.35   The output signal of a 64-bit organic RFID transponder chip with (a) optimized diode-load configuration and (b) optimized zero-V_{GS}-load configuration for a supply voltage of 20 V. The corresponding data rates are 4.3 kb/s for (a) and 522 b/s for (b).

because of limited pull-down current and the high parasitic output capacitance. The stage delay for both topologies increases with increasing back-gate voltages to the drive transistor. This effect can be attributed to a lower threshold voltage of the drive transistor yielding less current and therefore slower stage delay.

Next, we fabricated 64-bit RFID transponder chips [115] similar to earlier designs discussed in Section 5.2 [97]. To secure accurate clocking, we increased the size of the ring oscillator that generates the clock from 19 to 33 stages. For the optimized diode-load logic, the ratio between drive and load transistor is 10:1, leading to a transponder chip area of 74.48 mm². Figure 5.35 (a) shows the

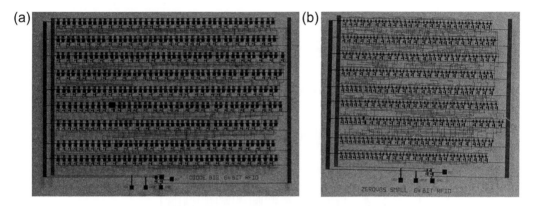

Figure 5.36   Photograph of the 64-bit organic RFID transponder chip designed with optimized diode-load configuration (a) and optimized zero-V_{GS}-load configuration (b).

output signal of the chip designed in optimized diode-load topology for a supply voltage of 20 V and $V_{BG,D}$ = 45 V. The data rate is 4.3 kb/s, more than twice the fastest single gate transponder chips shown in [97]. We verified on three different wafers of the same batch that some chips start operating at supply voltages as low as 10 V and all chips work at 15 V. The 64-bit transponder chip in optimized zero-V_{GS}-load topology is only 45.38 mm^2 in size, thanks to the 1:1 ratio used in this logic. In Figure 5.35 (b), the output is shown for a supply voltage of 20 V and $V_{BG,D}$ = 45 V. Also this transponder chip has been measured on three different wafers. All measured transponder chips are operational at 10 V. The data rate at supply voltage of 20 V is 522 b/s. These findings are fully in line with the fact that the zero-V_{GS}-load topology has higher noise margin but larger stage delay than the diode-load topology. Figure 5.36 shows a photograph of both 64-bit transponder chips. For comparison, a 64-bit transponder chip having single-gate zero-V_{GS}-load topology has also been fabricated on the same wafers. The onset voltage, more precisely the minimal supply voltage at which the transponder is functional, on the three wafers differs between 20 V and 26 V. These onset voltages for the power supply are comparable to the second generation transponder chips published in [97] and discussed in Section 5.3. As mentioned earlier, the dual-gate technology allows lowering of the supply voltage, even to 10 V, which is the lowest value reported to date for 64-bit transponder chips.

The back-gate voltages of the drive transistors are relatively high, 30 V to 45 V [115]. In an organic RFID transponder, such high voltage cannot be generated by a rectifier. However, it can be generated by charge pumps, provided that the required current is limited. To verify whether that last assumption is realistic, we have operated the chips with a 1-nA current compliance for the voltage supply of the back gate of the drive transistor. All chips were fully operational after a delay to charge all back gates. This delay was about 2.09 s for the diode-load configuration and only 246 ms for the zero-V_{GS}-load configuration. The output signal measured in the latter circuit is plotted in Figure 5.37. The difference in delay time is due to

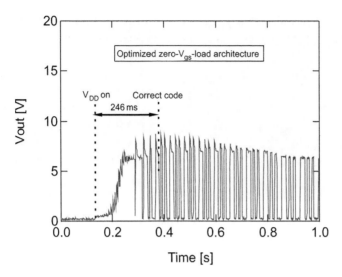

Figure 5.37   The output signal of an 8-bit organic RFID transponder chip with optimized zero-V_{GS}-load configuration for a supply voltage of 20 V and a back-gate voltage charged to 50 V at a current compliance of 1 nA.

the fact that the drive transistor is 10 times larger for the diode-load (ratioed) than the zero-V_{GS}-load configuration. We conclude that charge pumps are a viable route to charge the back gates of the logic blocks of the proposed dual-V_T technology. Recently, Marien et al. have demonstrated a DC-DC upconverter that can deliver biases between −40 V and 50 V at a supply voltage of 20 V to bias back gates of transistors in dual-gate circuits [127].

A better solution to enable lower back-gate voltage levels can be found in technology improvements targeting a higher oxide capacitance of the back-gate dielectric. This can be achieved by means of, for example, a thinner back-gate dielectric. A drawback of this solution is reduced performance in terms of stage delay of logic gates due to an increased output capacitor for the optimized dual-gate architecture.

5.7   Conclusions

In this chapter, we have presented the designs and technology evolutions that will allow thin-film electronics on plastic foil to fulfill the required specifications for EPC item-level tagging. In particular, we have shown that many of these specifications can be met. A 64-bit organic RFID tag has been elaborated; it is inductively coupled at a base carrier frequency of 13.56 MHz. The complexity of the chip has been increased to a 128-bit transponder chip. Also a basic ALOHA anti-collision protocol and WORM memory have been integrated. EPC-based data rates of 50 kb/s have been reached by downscaling of the technology. These data rates could also be met by using high-performing metal-oxide transistors or by the integration of diode-load or complementary logic. Recently, we demonstrated the possibility for plastic RFID tags to implement a bi-directional communication to realize an anti-collision protocol.

Among the specifications that have not been achieved yet, the most important is programmable memory arrays. Currently, research toward programmable memory arrays is ongoing. It will require a more complex design of the transponder chip. This implies that designs need more robustness against parameter variations of the technology. To improve the circuit robustness, we have proposed dual-V_T, unipolar technologies as illustrated by an integrated 64-bit RFID transponder chip and an 8-bit microprocessor (see the next chapter). Of course, the most viable option for increasing circuit robustness is to design and fabricate with complementary thin-film technologies, as discussed in this chapter, by the demonstration of a bi-directional communication chip.

6 Design Case: Organic Microprocessor

Forty years after the first silicon microprocessors, we have demonstrated an 8-bit microprocessor made by using a plastic electronic technology directly on flexible plastic foil [63]. The technology is based on optimized dual-gate zero-V_{GS}-load logic gates, making possible a gate level design flow as a result of its robustness, as discussed in Chapter 3. The operation speed of the microprocessor is today limited to 40 instructions per second. The power consumption is as low as 100 μW. The ALU-foil operates at a supply voltage of 10 V. The microprocessor can execute user-defined programs: we demonstrate the execution of the multiplication of two 4-bit numbers, and the calculation of the moving average of a string of incoming 6-bit numbers. To execute such dedicated tasks on the microprocessor, we create small plastic circuits that generate the sequences of appropriate instructions. The near-transparency, mechanical flexibility, and low power consumption of the processor are attractive features for integration on everyday objects, where it can be programmed as, among other applications, a calculator, timer, or game controller.

We have also realized a next generation thin-film microprocessor, based on the robust complementary technology [202]. This technology enabled the implementation of a more complex standard cell library, for example, a complex mirror adder cell. The number of instructions per second that can be executed is now 2100. Moreover, the dedicated instruction generator circuits have been replaced by a print-programmable instruction generator.

6.1 Introduction

Electronics pervades everyday life and is undeniably making its way from computing to telephony and increasingly continues to assist us in daily tasks through products such as electronic paper to read and write, electronic noses to sense gases, and smart lighting with electronics to save energy. The key enabler of these pervasive electronics applications is the fact that integration of ever more transistors with ever smaller dimensions has caused the cost of a single semiconductor transistor to dwindle to the level of 10 nano-dollars. Nevertheless, if the cost of a transistor in a chip is negligible and decreasing, the cost of placing and routing electronics on daily objects is not necessarily proportionally low.

Plastic electronics is the technology to make transistors and circuits with thin-film organic or plastic semiconductors on arbitrary substrates, including not only rigid substrates such as glass, but also flexible plastic foils. A variety of organic molecules and polymers have been developed as semiconductors, and the best ones [20]–[23] today feature a charge carrier mobility on the order of 1 to 10 cm^2/Vs, some 100- to 1000-fold lower than that of silicon. When integrated into circuits, the realistic mobility values are somewhat lower but nevertheless sufficient for applications such as back planes for flexible active-matrix displays, in particular for flexible electronic papers [94]. The first dedicated circuit applications of organic thin-film transistors have also appeared in recent years, such as recently demonstrated by the integration of an organic line driver for an organic active matrix OLED display [47]. Such circuits can be made directly on thin and ultra-flexible plastic foils, which allow them to be very simply laminated on everyday objects, and furthermore provide appealing characteristics in terms of bending radius and robustness: we no longer talk of flexible electronics but of truly crinkable electronics [202].

In this chapter, we investigate the possibility to use this technology to realize microprocessors on plastic foil [63]. As the cost of an electronic chip decreases with production volume, ultra-low cost microprocessors on easy-to-integrate flexible foils will be an enabler for ambient intelligence: one and the same type of chip can be integrated on vastly different types of objects to perform customized functions, such as identification, simple computing, and controlling.

The organic microprocessor has been implemented as two different foils: an arithmetic and logic unit (ALU) foil and an instruction foil. The ALU-foil is a general purpose foil that can execute a multitude of functions. On the other hand, the instruction foil is a dedicated chip that generates the sequence of instructions to obtain a specific function. It sends this sequence of instructions to the ALU-foil such that the combination of both foils results in the execution of a specific algorithm. The first prototype of the organic microprocessor [116] had only one instruction foil available and could operate up to 6 operations per second (OPS). In this work, we report an improved organic microprocessor that can run 40 OPS and can operate with two different instruction foils [63]. We first discuss the technology and choice of logic family used for the microprocessor foil. Subsequently, we report on the design and measurement data of the ALU-foil. Next, a complete integrated microprocessor is demonstrated by combining the ALU-foil with the instruction foil. Finally, we conclude by comparing the organic microprocessor to the silicon Intel 4004 early days processor.

6.2   Technology and Logic Family

The technology used in this work is the organic, dual-gate, thin-film transistor technology from Polymer Vision [95] and discussed in more detail in Chapter 2, Section 2.4.2. As concluded from Chapter 3, dual-gate transistors can be integrated into

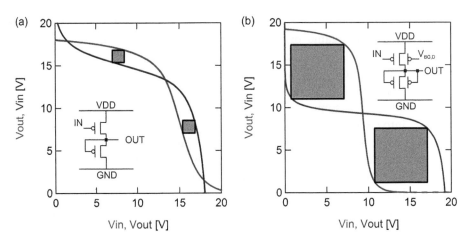

Figure 6.1   VTC and noise margin analysis (a) of a single-V_T, zero-V_{GS}-load inverter with single-gate transistor technology and (b) of an optimized dual-V_T inverter using dual-gate transistor technology.

unipolar, dual-V_T logic gates yielding an increased circuit robustness compared to single-V_T logic gates [115]. Figure 6.1 shows the noise margin (at $VDD = 20$ V) of typical zero-V_{GS}-load inverters when no back gate is used, compared to the noise margin achievable with the optimized dual-gate zero-V_{GS}-load topology. In this optimized topology, the back gates of the load transistors are connected to the front gates, while all back gates of the drive transistors are connected to a common rail, to which a back-gate voltage is applied externally [115]. For more details on these logic gates, we like to refer to Chapter 3.

The typical spread on threshold voltage in organic TFT technology is 0.2 to 0.5 V, large compared to the noise margin achievable with single-gate technology. As a result, it is common practice in the field of thin-film electronics to use a transistor-level approach to design (simple) circuits. Indeed, it is usually necessary to simulate the schematic entry with an analog circuit level simulator (such as Spectre or Spice) and use Monte Carlo simulations to predict yield, as explained in Chapter 4 and demonstrated with the design case RFID in the previous chapter. However, such analog circuit level simulators are not adapted to deal with the required level of complexity to design and simulate an organic microprocessor because of the large number of (parallel) switching gates and large amount of input, control, and output signals. In contrast, using our optimized dual-gate logic gates, the much improved noise margin allows the use of common digital design practices. Starting from the basic characteristics of inverters and other logic gates, we designed a robust library of basic digital logic gates (inverters, NANDs, buffers). This standard cell library was used to design the organic microprocessor by means of a gate-level design approach. Therefore, after modeling, simulating, and measuring the basic building blocks, we used a gate-level simulator (Modelsim) with our standard cell library to design and simulate the organic microprocessor. The

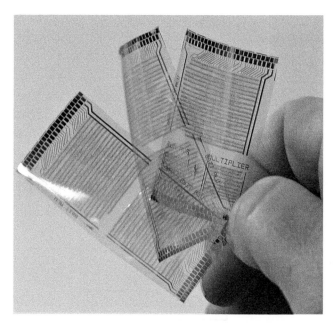

Figure 6.2   Photograph of some microprocessor foils: the leftmost foils each have two microprocessors; the rightmost foil has a microprocessor and two dedicated instruction foils. Each microprocessor is about $2 \times 1.7\text{cm}^2$ and contains 3381 plastic transistors.

W/L ratio between drive and load transistor for the logic gates in the library was a 1:1 ratio beneficial for area, with a minimal W/L of 140/5 μm/μm (as discussed in Chapter 3) [63], [115].

Figures 6.2 and 6.11 show a photograph of some microprocessors on foil. The 25-micrometer-thick foil is highly flexible. Furthermore, the complete circuit is nearly transparent, as only the metal electrodes of gates, sources, drains, and interconnection lines are reflective.

6.3   Architecture and Measurement Results of the Organic ALU-Foil

Characteristic of a microprocessor is that its hardware is not dedicated to a single function or operation, but is designed such that the operations performed on (digital) inputs can be programmed and defined after manufacture of the processor. The challenge, therefore, is to manage the plurality of possible critical data paths in the microprocessor, for all different instruction codes and inevitable variations due to the nature of organic technology on foil. Our microprocessor has been constructed around an 8-bit arithmetic and logic unit (ALU), which comprises three blocks as schematically represented in Figure 6.3. The first block adds or subtracts the incoming numbers (arithmetic unit), the second block performs logic operations on the incoming data (logic unit), and the third block shifts the incoming

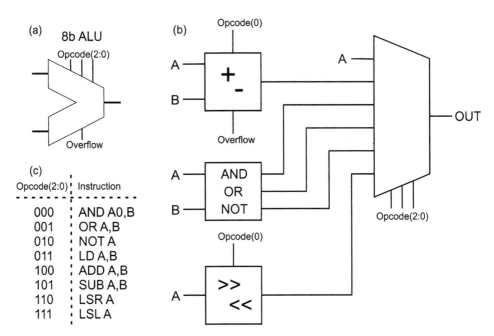

Figure 6.3   (a) Symbol, (c) instruction set, and (b) architecture of the main building block of the microprocessor, namely, the 8-bit ALU. Three operational code bits, opcode (2:0), are used to select among the different instructions.

bits (bit shift unit). Figure 6.4 details the design choices for each of these units. The arithmetic unit is designed as a ripple carry adder/subtractor, as schematically depicted in Figure 6.4 (a). The logic unit can execute the AND, OR and NOT operations. In order to ease the implementation of a multiplier, we have chosen to design the AND as a logic AND of bit A0 with B, shown in Figure 6.4 (b). The bit shift units are implemented as logic shifts (LSR, LSL). This implies that, after the shift operation of the word, a '0' is included at the void, schematically represented in Figure 6.4 (c). Detailed control over each of these three blocks and the actual selection of the output of the ALU unit is determined by the microprocessor's instruction set, also called operational codes or "opcodes" (Figure 6.3). As the architecture depicts, the ALU executes every clock cycle instruction on each of the three units in parallel. Subsequently, a multiplexer selects the desired instruction to be executed in that clock cycle.

Figure 6.5 outlines the complete architecture of the microprocessor foil. Around this ALU, a minimal set of 8-bit registers has been placed, for storing the working data (accumulator A, working registers [C_0, C_1, and C_2], and an output register). The storing and loading of the data in these registers are also controlled by the instruction set (see Table 6.1). The registers select bits (RR in Table 6.1) that correspond to bits 7 and 8 of the opcode and are used to select among the four working registers, C_0 to C_3. Working register C_3 is implemented as a hard-coded decimal 1 in order to ease the implementation of the increment and decrement instructions.

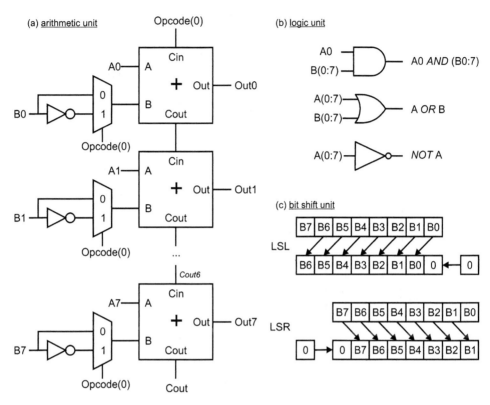

Figure 6.4 Detailed zoom of the design choice for the (a) arithmetic unit, (b) logic unit, and (c) bit shift unit.

We have tested all the individual instructions of the microprocessor foil extensively for different bias conditions using a dedicated hardware testbench running on a PIC18F development board. The expected outcome (generated by the PIC18F testboard) and the measured output register value are plotted on a digital oscilloscope (see Figure 6.6). As such, the correct behavior of the microprocessor can be easily verified.

Figure 6.7 (a) shows that the microprocessor can perform up to 40 operations per second (OPS), when powered at 20 V supply voltage and an appropriate back-gate voltage. This maximum frequency is determined by the 25 ms critical path delay in the design. Figure 6.7 (b) shows that the microprocessor can operate at voltages down to 10 V. The critical path is defined by the subtract operation, which requires the carry bits to ripple through all eight full adders. As shown in Figure 6.7 (c), the contribution of these logic gates was measured separately on different kinds of ring oscillators, as a function of the capacitive load of the gates. An inverter driving a single subsequent stage has a minimum capacitive load, and in that case its gate delay is 83 µs. Similarly, the minimum gate delay of a two-input NAND is 126 µs, while one input is connected to *VDD*. However, when a logic gate has to drive multiple subsequent stages in parallel, it is slowed down: we

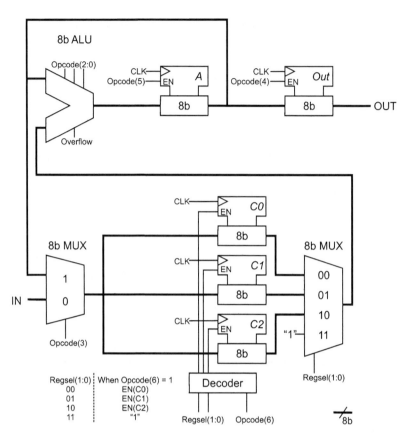

Figure 6.5   Architecture of the microprocessor core, comprising the arithmetic and logical unit (ALU), accumulator register "A," and output register "OUT" at the top and the input multiplexer and storage registers "C" at the bottom.

show in Figure 6.7 (c) that a gate driving nine identical inverter gates in parallel is slowed down to 1 ms. This gate delay, combined with the length of the critical path, explains why with our current design and topology, the processor frequency is 40 OPS. Moreover, because it was the first time a circuit of this complexity was realized in organic technology, we preferred conservative design choices. For instance, we utilized only gates with a fan-in of 2 and our signal buffering strategy was very conservative. By alleviating these restrictions and by optimizing the design, we estimate that the frequency can improve to the hundreds of OPS range. Another reason for the current limitation to the tens of OPS range is related to the choice of logic family, where we have chosen for robustness. Other unipolar logic types (dual-gate, diode-connected) are more advantageous in terms of speed [115] (see Chapters 3 and 5, Section 5.6). As Figure 6.7 also depicts, the frequency of the ALU and the ring oscillator decreases when the back-gate voltage of the drive transistors increases. This can be explained by a negative V_T-shift of the drive TFT yielding less drive current [115].

Table 6.1. Implemented instruction set: RR refers to the binary representation of the selected register number; X is a "don't care"

Opcode (9:0)	Instruction	Function
0RR0100000	AND A(0), C$_{RR}$	Logic
0RR0100001	OR A, C$_{RR}$	Logic
0XX0100010	NOT A	Logic
0RR0100011	LD A, C$_{RR}$	Load
0RR0100100	ADD A, C$_{RR}$	Arithmetic
0RR0100101	SUB A, C$_{RR}$	Arithmetic
0XX0100110	LSR A	Bit shift
0XX0100111	LSL A	Bit shift
0XX000XXXX	NOOP	No operation
0RR1001XXX	LD C$_{RR}$, A	Load
0RR1000XXX	LD C$_{RR}$, IN	Load
0XX001XXXX	LD OUT, A	Load
0110100100	INC A	Arithmetic
0110100101	DEC A	Arithmetic
10000AAAAA	JMP AAAAA	Jump

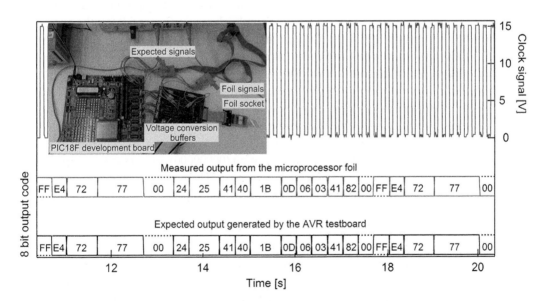

Figure 6.6 Hardware testbench measuring all individual instructions of the microprocessor foil at a clock frequency of 6 Hz (V_{Back} = 50 V and VDD = 15 V). The inset shows a photograph of the measurement setup.

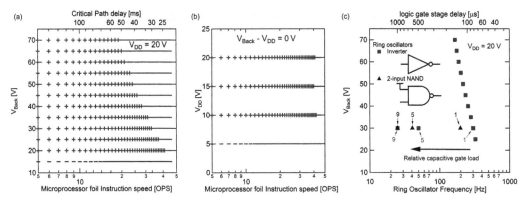

Figure 6.7   Shmoo plot of the microprocessor core as a function of the back-gate voltage (a) and of the power voltage (b): the highest frequency is reached at the lowest back-gate voltage (V_{Back}) that still allows correct operation. The top axis shows the corresponding critical path delay. (c) Measurement of the operation speed of building blocks: squares: ring oscillator frequency (bottom axis) versus V_{Back} (left axis) of a 19-stage ring oscillator, measured to explain the processor frequency. The top axis shows the gate delay of the individual inverters of the chain. The stage delay of inverters driving five and nine gates has corresponding labels. The triangles show the measured stage delay of two-input NAND gates driving one, five, and nine subsequent gates.

Furthermore, we show that our microprocessor is truly a general-purpose machine that can be programmed for multiple uses by executing instruction codes for different applications. In a first example, shown in Figure 6.8 (a), the microprocessor is programmed to execute a multiplication of two numbers. The solid line shows a sequence of cycles, in each of which two 4-bit numbers are multiplied to give an 8-bit output value, shown as a dashed line. The input values are shown on the left scale (from 0 to 15), while the output is shown on the right scale (0 to 255). The first cycle shows the multiplication of the binary numbers 0111 (decimal 7) and 1010 (decimal 10) to 01000110 (decimal 70). This value remains at the output while two new numbers are fed in at the input (0000 and 1111) and multiplied. In the third cycle, the output of this multiplication (00000000) appears at the output while a new set is provided and executed, and so on. The processor clock can be verified to run at 40,Hz during this execution, as explained previously. Table 6.2 details the instruction set applied to the microprocessor by the test board. It also shows an example of the multiplication of 1010 by 0111, including the list of instructions needed to acquire intermediate results.

In a different example, chosen from the application domain of digital signal processing, the microprocessor executes the weighed time-averaging of a stream of incoming digital inputs to reduce random noise. This algorithm is known to clean up the output signal of a sensor after digitization by an analog-to-digital converter (ADC). By virtue of its applicability to large area substrates, plastic electronics is suited to develop large area sensors [60], and the first plastic ADC converters were shown recently [61], [123]. We implemented the algorithm of a moving averager

Table 6.2. Program listing of the dedicated instruction set of the multiplier with an example of the multiplication of 1010 and 0111

PC	Instruction	PC	Instruction	PC	Instruction	PC	Instruction
1	LD C_{00}, IN	9	LD A, C_{00}	17	LD C_{01}, A	25	LSL A
2	LD C_{01}, IN	10	LSR A	18	LD A, C_{00}	26	LD C_{01}, A
3	LD A, C_{00}	11	LD C_{00}, A	19	LSR A	27	LD A, C_{00}
4	AND A(0), C_{01}	12	AND A(0), C_{01}	20	LD C_{00}, A	28	LSR A
5	LD C_{10}, A	13	ADD A, C_{10}	21	AND A(0), C_{01}	29	LD C_{00}, A
6	LD A, C_{01}	14	LD C_{10}, A	22	ADD A, C_{10}	30	AND A(0), C_{01}
7	LSL A	15	LD A, C_{01}	23	LD C_{10}, A	31	ADD A, C_{10}
8	LD C_{01}, A	16	LSL A	24	LD A, C_{01}	32	LD OUT, A

				1	0	1	0	*10*	Reg C_{01}
				0	1	1	1	*x 7*	Reg C_{00}
				1	0	1	0		PC #3–5
			1	0	1	0	0		PC #6–14
		1	0	1	0	0	0		PC #15–23
	0	0	0	0	0	0	0		PC #24–31
0	1	0	0	0	1	1	0	*70*	

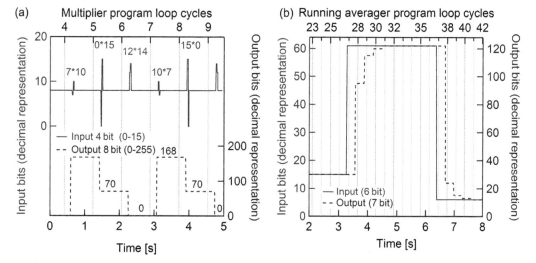

Figure 6.8 Demonstration of user-programmability of the microprocessor. (a) Time evolution of the 4-bit input signals and the 8-bit output signal when the multiplier instruction set (see Table 6.2) is implemented in the organic microprocessor foil. The top axis shows the cycle number of the program loop. (b) The 6-bit input signal (left axis) and the 7-bit output signal (right axis, which has one significant binary digit more than the input) when the running averager program (see Table 6.3) is executed. The top axis shows the cycle number of the program loop.

Table 6.3. Program listing of the dedicated instruction set of the running averager

PC	Instruction	PC	Instruction
1	LD A, C_{03}	9	LD C_{01}, A
2	SUB A, C_{03}	10	LD C_{00}, IN
3	LD C_{01}, A	11	LD A, C_{00}
4	LD C_{00}, IN	12	ADD A, C_{01}
5	LD A, C_{00}	13	LD OUT, A
6	ADD A, C_{01}	14	LSR A
7	INC A	15	JUMP #3
8	LSR A	16	NOOP

$out_{new} = 0.5$ round ($in + out_{old}$), that is, an averaging algorithm in which the weight of the past data decreases exponentially, and demonstrate the execution of the algorithm in Figure 6.8 (b). The 6-bit input provided to the microprocessor is shown as the solid line: 001111 (15) during the first 4 loop cycles, then 111101 (61) during the next 10 cycles, then 000110 (6). The running averager calculates the weighed average as a 7-bit number, which can be seen to tend to the steady input values after they have been provided for some cycles. Here again, the clock speed of the processor is 40 Hz. Table 6.3 details the instruction set. Instruction addresses #1 and #2 are used as reset; instructions #3 to #14 are the execution of the averaging algorithm. In order to bypass the reset action at the end of the instructions, we added instruction #15, which jumps to address #3 to form a loop. The algorithm is executed twice each program loop and the input is also sampled twice each program loop. In this algorithm, we have included once an INC operation prior to the LSR instruction in order to make sure the output can converge to the inputs' value. The output pins are only updated after the second implementation, prior to the LSR A instruction. This implies that the output register will contain a 7-bit number, which will finally average out – in decimal representation – to twice the input value.

6.4   Integrated Organic Microprocessor on Foil

Until now only the ALU-foil of the microprocessor core was discussed. In the preceding demonstrations, the instructions for the microprocessor were generated by external test equipment. To arrive at a complete plastic solution we also designed a plastic control unit, shown in Figure 6.9. This control unit has as its task to take instructions from a memory in the appropriate order controlled by a program counter. These instructions are sent as opcodes to the microprocessor core. Opcodes for the program counter are also generated to enable branching in the program. In the ideal case the program would be stored in programmable, non-volatile memory on the foil. However, programmable non-volatile memory compatible with plastic thin-film

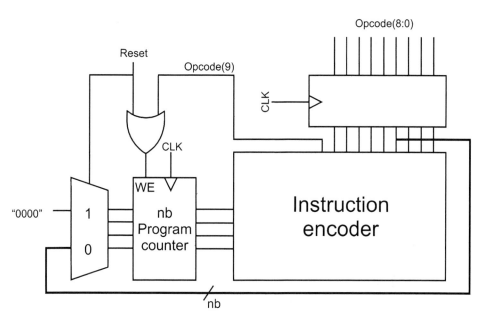

Figure 6.9   Block diagram of the instruction generating foils: n is 4 in the case of the averager foil.

transistors on foil is still subject of research today and is as such not available for our experiments. Therefore, as in the early days of silicon technology, we used true read only memory (ROM): the instructions are hardcoded on the foil. A different foil is designed for every program. For the low cost, low complexity, but high volume applications that are envisaged here, this procedure could even be a realistic commercial scenario. The instruction sets generated by the multiplier instruction foil and the moving averager foil are shown in Tables 6.2 and 6.3, respectively.

The operation of the running averager instruction foil by itself is shown in Figure 6.10 (a). This circuit does not contain a ripple carry adder, and therefore it has a shorter critical path compared to the microprocessor. Stand-alone, the instruction circuit can run at a clock speed of 70 Hz. The photograph of the running averager foil is shown in Figure 6.10 (b).

Finally, we demonstrate the combined operation of an instruction foil with the microprocessor. We conducted this experiment with the running averager. The correct operation of this combination is shown in Figure 6.10 (c). This demonstration shows that the microprocessor can indeed accept its instruction set from a dedicated plastic circuit and is not limited to instruction sets from a test board.

6.5   Second Generation Thin-Film Processor

The second generation thin-film microprocessor [201] has been realized in the complementary technology, described in Chapter 2, Section 2.4.5. This complementary

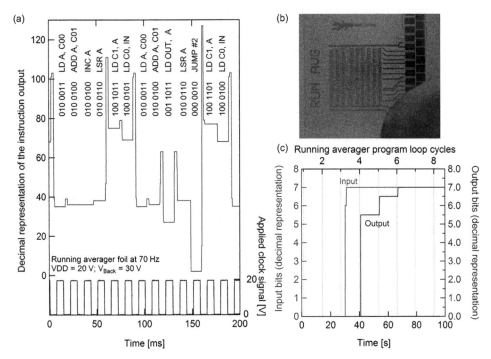

Figure 6.10   Demonstration of operation of the plastic microprocessor commanded by an instruction foil. (a) Decimal representation of the measured seven least significant bits of the instruction generated by the running averager instruction foil running at 70 Hz. The clock is shown at the right-hand axis. The data are valid on the rising edge of the clock. (b) Photograph of the running averager foil. (c) measured output of the microprocessor foil connected to the running averager instruction foil. As the input bit stream switches from 000000 to 000111, the output gradually increases to the same level over three loop cycles, but with seven significant binary digits.

technology increases the robustness of standard cells and therefore the soft yield of the final circuit. As a consequence, we have designed a more complex and quasi-complete standard cell library, with inverters, NAND and NOR gates, buffer cells, and a complex mirror adder cell. The data path of the processor has been improved by exchanging the previously combinatorial implementation of the full adder with the mirror adder cell and by appropriate buffering of the internal signals. This leads to a minimal transistor count of the data path, lowering the area consumption. Aside from the improvements on the data paths and the adequate signal buffering, the general processor architecture is the same as depicted in Figure 6.5. As a consequence of these technological and design improvements, this second generation processor core chip or ALU chip can execute maximum 2100 operations/ second at 12 V supply voltage. The chip picture and measured OPS versus supply voltage are shown in Figure 6.12. The chip is operational from 6.5 V onward.

The processor core chip also requires a small chip that provides the necessary instructions to execute a specific function. The second generation instruction

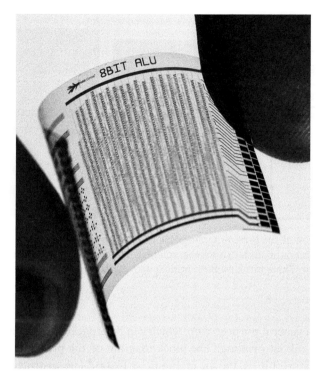

Figure 6.11 Photograph of the 8 bit ALU-foil.

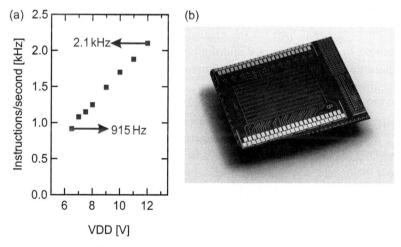

Figure 6.12 (a) Measured maximum number of instructions per second as a function of the chip's supply voltage and (b) die photograph of the 8 bit ALU chip.

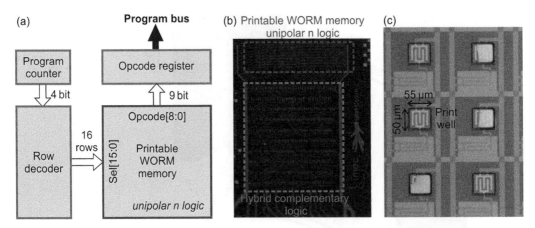

Figure 6.13   (a) The block diagram of the print-programmable instruction generator, (b) die picture of the instruction generator, and (c) a micrograph image of the inkjet printed area of the instruction generator. Three bits have been printed with Ag ink to logic 0s.

generator chip is a general purpose chip, which can be programmed for a dedicated task by means of ink jet printing. The block diagram of the print-programmable instruction generator chip is shown in Figure 6.13 (a). The instruction generator features a 4-bit program counter, which uses the same clock as the data path. A 16-line address decoder selects a row in the printable write-once read-many times (WORM) memory, and a 9-bit register subsequently stores the selected instruction and provides it on the program bus. Instructions can be programmed in the memory via a simple inkjet printing step. The layout of this configurable instruction generator chip is depicted in Figure 6.13 (b). Figure 6.13 (c) shows an expanded view of the layout. The area for the printed ink drop is 50 μm × 55 μm. The lines to be shorted by the ink drop are interdigitated, with 5 micrometer spacing. This design guarantees 100 percent reliable shorting, even when inkjet equipment of modest accuracy is used.

The printable WORM memory block is implemented as a 16-input unipolar NOR gate as shown in Figure 6.14 (a). The load transistor is connected as depletion-load. In other words, the gate is connected to its source. This unipolar topology was selected because the TFTs exhibit a normally-on behavior. Up to 16 drive transistors (Sel0, Sel1, …) can be connected to the NOR gate by inkjet printing a droplet in the well. Figure 6.14 (b) plots the simulated voltage transfer curves of the NOR gate for varying numbers of inputs. These simulations were performed by sweeping the selected drive transistor, in this case, Sel0 (= V_{in}), between 0 V and 10 V while biasing all remaining unselected transistors with 0 V V_{GS}.

Increasing the number of drive transistors decreases the noise margin, as a result of the additional leakage from the unselected TFTs at a gate-source voltage bias of 0 V. To compensate for this, additional load transistors can be added through inkjet printed connections. The simulated voltage transfer characteristics

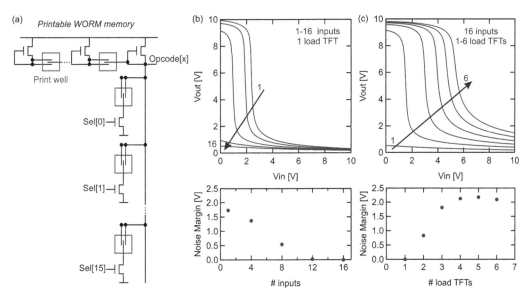

Figure 6.14   (a) The transistor schematic of the printable WORM memory. The voltage transfer characteristics and the extracted noise margins are plotted for (b) a 1–16 input unipolar NOR gate with only one load transistor and (c) a 16 input NOR gate with increased number of load transistors 1–6.

of the 16-input NOR gate with increasing number of load transistors are shown in Figure 6.14 (c). The noise margin can be recovered when more load transistors are added. For this configuration, the optimal balance is observed to be five load TFTs for 16 inputs. This optimum ratio changes as the threshold voltage of the oxide TFTs varies, for example, as a result of process corners or variability. Another optimum is found when the bias voltage at the gates of the unselected lines is larger than 0 V, in case the complementary logic is not fully rail-to-rail. Even though the printable WORM memory is designed in a unipolar technology, it still offers good robustness because of the ability to add multiple load transistors in post-processing.

In total, the print-programmable instruction generator employs 403 organic p-TFTs and 412 oxide n-TFTs in its unprogrammed state and measures 9.0 mm × 6.9 mm. Printing the connections for the drive and load transistors in the unipolar oxide NOR gate can add up to 189 oxide n-TFTs. This number will differ according to the desired program and routines.

The thin-film P²ROM chip can store 16 lines of 9-bit instructions. As an example, we have implemented an exponential running averager algorithm configured by inkjet printing the P²ROM chip. The algorithm is executed twice in one cycle, and the output register is enabled after the second loop, prior to the second LSR function. This increases the accuracy with 1 bit. The algorithm is executed by 12 subsequent instructions. The remaining 4 available commands are configured as NOOP (no operation). Figure 6.15 (a) plots the measured outputs of the P²ROM chip at

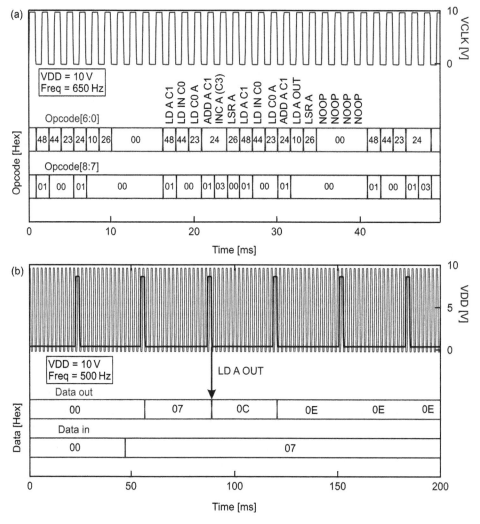

Figure 6.15 (a) Measured signals of the P²ROM instruction generator when configured (printed) to execute the running averager algorithm. It consists of 12 instructions and 4 NOOP commands. (b) Measured signals of both the P²ROM and processor core chips while executing a running averager algorithm. The pulses in the top part of the figure correspond to the command "store in output register."

a supply voltage of 10 V and clock speed of 650 Hz. The corresponding instructions are shown in the graph. Configured as a first order averager, the P²ROM chip employs 852 TFTs. Figure 6.15 (b) shows the measurement results for the full computer when the processor core and P²ROM chips are combined. The input stream changes from 0 (000000, 6 bit) to 7 (000111, 6 bit). The 7-bit accurate outputs are subsequently 0, 7, C, and E, in hexadecimal numbers. The graph also details the instruction from the P²ROM chip, which enables the output register.

Table 6.4. Comparison with the early silicon processor

	Plastic microprocessor	Second generation thin-film microprocessor	Intel 4004
Transistor-count processor core	3381	3504	2300
Transistor-count instruction generator	612 (averager)	852 (P²ROM averager)	
Area	1.96×1.72 cm²	1.20×1.88 cm²	3×4 mm²
Pin-count	30	30	16
Min. power supply voltage	10 V	6.5 V	15 V
Power consumption	92 μW	Not available	1 W
Operation speed	40 operations/second	2100 operations/second	92000 operations/second
Semiconductor	Pentacene	Pentacene (p-TFT) and sol. processed oxide (n-TFT)	Silicon
P-type mobility	~0.15 cm²/Vs	0.15 cm²/Vs (p-TFT) 2.00 cm²/Vs (n-TFT)	~450 cm²/Vs
Logic family	p-type only	p + n	p-type only
Operation	Accumulation	Accumulation	Inversion
Technology	5 μm	5 μm	10 μm
Bus width	8 bit	8 bit	4 bit
Demonstration year	2011	2014	1971
Wafer scale	6"	6"	2"
Substrate	Flexible	Rigid (Flexible possible)	Rigid

6.6   Conclusions

To conclude, we compare in Table 6.4 the characteristics of the first and second generation thin-film microprocessor with the early silicon processors made in p-type-only silicon technology some four decades ago [203]. Significant correspondence can be seen regarding parameters such as gate length, supply voltage, and transistor count, but some marked differences are also clear. The instruction rate of the organic technology is about three orders of magnitude slower than the early silicon processor, as a direct consequence of the three order of magnitude lower carrier mobility in organic semiconductors. However, on the positive side, the power consumption is also four orders of magnitude smaller. The second generation thin-film microprocessor has an increased operation speed, which is now only 44 times less compared to the Intel 4004. Moreover, the minimal supply voltage has been lowered to 6.5 V. The root cause for these improvements is the complementary technology. This robust technology enabled a quasi-complete standard cell library used to optimize the data path of the processor core chip. Moreover, the second generation chipset is a general purpose design, which can be programmed using a post-manufacturing digital print step.

Bibliography

[1] R. A. Street, "Thin-film transistors," *Advanced Materials*, vol. 21, no. 20, pp. 2007–2022, 2009.

[2] S. Wagner, H. Gleskova, I.-C. Cheng, and M. Wu, "Silicon for thin-film transistors," *Thin Solid Films*, vol. 430, no. 1–2, pp. 15–19, Apr. 2003.

[3] M. J. Powell, "Charge trapping instabilities in amorphous silicon-silicon nitride thin-film transistors," *Applied Physics Letters*, vol. 43, no. 6, pp. 597–599, Sep. 1983.

[4] M. J. Powell, C. van Berkel, and J. R. Hughes, "Time and temperature dependence of instability mechanisms in amorphous silicon thin-film transistors," *Applied Physics Letters*, vol. 54, no. 14, pp. 1323–1325, Apr. 1989.

[5] F. R. Libsch and J. Kanicki, "Bias-stress-induced stretched-exponential time dependence of charge injection and trapping in amorphous thin-film transistors," *Applied Physics Letters*, vol. 62, no. 11, pp. 1286–1288, Mar. 1993.

[6] A. Nathan, G. R. Chaji, and S. J. Ashtiani, "Driving schemes for a-Si and LTPS AMOLED displays," *Journal of Display Technology*, vol. 1, no. 2, pp. 267–277, Dec. 2005.

[7] Y. He, R. Hattori, and J. Kanicki, "Improved a-Si:H TFT pixel electrode circuits for active-matrix organic light emitting displays," *IEEE Transactions on Electron Devices*, vol. 48, no. 7, pp. 1322–1325, Jul. 2001.

[8] J.-C. Goh, J. Jang, K.-S. Cho, and C.-K. Kim, "A new a-Si:H thin-film transistor pixel circuit for active-matrix organic light-emitting diodes," *IEEE Electron Device Letters*, vol. 24, no. 9, pp. 583–585, Sep. 2003.

[9] J.-H. Lee, J.-H. Kim, and M.-K. Han, "A new a-Si:H TFT pixel circuit compensating the threshold voltage shift of a-Si:H TFT and OLED for active matrix OLED," *IEEE Electron Device Letters*, vol. 26, no. 12, pp. 897–899, Dec. 2005.

[10] A. Kumar, A. Nathan, and G. E. Jabbour, "Does TFT Mobility Impact Pixel Size in AMOLED Backplanes?" *IEEE Transactions on Electron Devices*, vol. 52, no. 11, pp. 2386–2394, Nov. 2005.

[11] Y. Yamamoto, "Technological innovation of thin-film transistors: Technology development, history, and future," *Japanese Journal of Applied Physics*, vol. 51, p. 060001, 2012.

[12] S. D. Brotherton, "Polycrystalline silicon thin film transistors," *Semiconductor Science and Technology*, vol. 10, no. 6, pp. 721–738, Jun. 1995.

[13] J. B. Choi, Y. J. Chang, C. H. Park, B. R. Choi, H. S. Kim, and K. C. Park, "TFT backplane technologies for AMLCD and AMOLED applications," *Journal of the Korean Physical Society*, vol. 54, no. 925, p. 549, Jan. 2009.

[14] S. Uchikoga and N. Ibaraki, "Low temperature poly-Si TFT-LCD by excimer laser anneal," *Thin Solid Films*, vol. 383, no. 1–2, pp. 19–24, Feb. 2001.

[15] J.-I. Ohwada, M. Takabatake, Y. A. Ono, A. Mimura, K. Ono, and N. Konishi, "Peripheral circuit integrated poly-Si TFT LCD with gray scale representation," *IEEE Transactions on Electron Devices*, vol. 36, no. 9, pp. 1923–1928, Sep. 1989.

[16] J. Y. Yang, S.-H. Kim, Y.-I. Park, K.-M. Lim, and C.-D. Kim, "P-2: A novel structure of AMLCD panel using poly-Si CMOS TFT," *SID Symposium Digest of Technical Papers*, vol. 35, no. 1, pp. 224–227, 2004.

[17] J.-C. Goh, H.-J. Chung, J. Jang, and C.-H. Han, "A new pixel circuit for active matrix organic light emitting diodes," *IEEE Electron Device Letters*, vol. 23, no. 9, pp. 544–546, Sep. 2002.

[18] C.-L. Fan, M.-C. Shang, W.-C. Lin, H.-C. Chang, K.-C. Chao, and B.-L. Guo, "LTPS-TFT pixel circuit compensating for TFT threshold voltage shift and IR-drop on the power line for AMOLED displays," *Advances in Materials Science and Engineering*, vol. 2012, pp. 1–5, 2012.

[19] H. Klauk, "Organic thin-film transistors," *Chemical Society Reviews*, vol. 39, no. 7, pp. 2643–2666, 2010.

[20] P. T. Herwig and K. Müllen, "A Soluble Pentacene Precursor: Synthesis, Solid-State Conversion into Pentacene and Application in a Field-Effect Transistor," *Advanced Materials*, vol. 11, no. 6, pp. 480–483, 1999.

[21] J. Chen, S. Subramanian, S. R. Parkin, M. Siegler, K. Gallup, C. Haughn, D. C. Martin, and J. E. Anthony, "The influence of side chains on the structures and properties of functionalized pentacenes," *Journal of Materials Chemistry*, vol. 18, no. 17, p. 1961, 2008.

[22] N. Kobayashi, M. Sasaki, and K. Nomoto, "Stable peri-Xanthenoxanthene thin-film transistors with efficient carrier injection," *Chemistry of Materials*, vol. 21, no. 3, pp. 552–556, Feb. 2009.

[23] B. Yoo, B. A. Jones, D. Basu, D. Fine, T. Jung, S. Mohapatra, A. Facchetti, K. Dimmler, M. R. Wasielewski, T. J. Marks, and A. Dodabalapur, "High-performance solution-deposited n-channel organic transistors and their complementary circuits," *Advanced Materials*, vol. 19, no. 22, pp. 4028–4032, 2007.

[24] M. J. Kang, I. Doi, H. Mori, E. Miyazaki, K. Takimiya, M. Ikeda, and H. Kuwabara, "Alkylated Dinaphtho[2,3-b:2′,3′-f]Thieno[3,2-b]Thiophenes (Cn-DNTTs): Organic semiconductors for high-performance thin-film transistors," *Advanced Materials*, vol. 23, no. 10, pp. 1222–1225, 2011.

[25] R. R. Lunt, B. E. Lassiter, J. B. Benziger, and S. R. Forrest, "Organic vapor phase deposition for the growth of large area organic electronic devices," *Applied Physics Letters*, vol. 95, no. 23, pp. 233305–233305–3, Dec. 2009.

[26] C. Rolin, S. Steudel, P. Vicca, J. Genoe, and P. Heremans, "Functional pentacene thin films grown by in-line organic vapor phase deposition at web speeds above 2 m/min," *Applied Physics Express*, vol. 2, no. 8, p. 086503, 2009.

[27] C. Rolin, K. Vasseur, B. Niesen, M. Willegems, R. Müller, S. Steudel, J. Genoe, and P. Heremans, "Vapor phase growth of functional pentacene films at atmospheric pressure," *Advanced Functional Materials*, vol. 22, no. 23, pp. 5050–5059, 2012.

[28] G. H. Gelinck, T. C. T. Geuns, and D. M. de Leeuw, "High-performance all-polymer integrated circuits," *Applied Physics Letters*, vol. 77, no. 10, pp. 1487–1489, Sep. 2000.

[29] Y.-Y. Noh, N. Zhao, M. Caironi, and H. Sirringhaus, "Downscaling of self-aligned, all-printed polymer thin-film transistors," *Nature Nanotechnology*, vol. 2, no. 12, pp. 784–789, 2007.

[30] M. Hambsch, K. Reuter, M. Stanel, G. Schmidt, H. Kempa, U. Fügmann, U. Hahn, and A. C. Hübler, "Uniformity of fully gravure printed organic field-effect transistors," *Materials Science and Engineering: B*, vol. 170, no. 1–3, pp. 93–98, Jun. 2010.

[31] A. Daami, C. Bory, M. Benwadih, S. Jacob, R. Gwoziecki, I. Chartier, R. Coppard, C. Serbutoviez, L. Maddiona, E. Fontana, and A. Scuderi, "Fully printed organic CMOS technology on plastic substrates for digital and analog applications," in *IEEE International Solid-State Circuits Conference (ISSCC)*, 2011, pp. 328–330.

[32] A. Shin, S. J. Hwang, S. W. Yu, and M. Y. Sung, "Design of organic TFT pixel electrode circuit for active-matrix OLED displays," *Journal of Computers*, vol. 3, no. 3, pp. 1–5, 2008.

[33] V. Vaidya, S. Soggs, J. Kim, A. Haldi, J. N. Haddock, B. Kippelen, and D. M. Wilson, "Comparison of pentacene and amorphous silicon AMOLED display driver circuits," *IEEE Transactions on Circuits and Systems I: Regular Papers*, vol. 55, no. 5, pp. 1177–1184, Jun. 2008.

[34] P.-T. Liu and L.-W. Chu, "Innovative voltage driving pixel circuit using organic thin-film transistor for AMOLEDs," *Journal of Display Technology*, vol. 5, no. 6, pp. 224–227, Jun. 2009.

[35] K. Nomura, H. Ohta, K. Ueda, T. Kamiya, M. Hirano, and H. Hosono, "Thin-film transistor fabricated in single-crystalline transparent oxide semiconductor," *Science*, vol. 300, no. 5623, pp. 1269–1272, May 2003.

[36] K. Nomura, H. Ohta, A. Takagi, T. Kamiya, M. Hirano, and H. Hosono, "Room-temperature fabrication of transparent flexible thin-film transistors using amorphous oxide semiconductors," *Nature*, vol. 432, no. 7016, pp. 488–492, Nov. 2004.

[37] E. Fortunato, P. Barquinha, and R. Martins, "Oxide semiconductor thin-film transistors: A review of recent advances," *Advanced Materials*, vol. 24, no. 22, pp. 2945–2986, 2012.

[38] J. S. Park, W.-J. Maeng, H.-S. Kim, and J.-S. Park, "Review of recent developments in amorphous oxide semiconductor thin-film transistor devices," *Thin Solid Films*, vol. 520, no. 6, pp. 1679–1693, Jan. 2012.

[39] J. K. Jeong, J. H. Jeong, J. H. Choi, J. S. Im, S. H. Kim, H. W. Yang, K. N. Kang, K. S. Kim, T. K. Ahn, H.-J. Chung, M. Kim, B. S. Gu, J.-S. Park, Y.-G. Mo, H. D. Kim, and H. K. Chung, "3.1: Distinguished paper: 12.1-Inch WXGA AMOLED display driven by indium-gallium-zinc oxide TFTs array," *SID Symposium Digest of Technical Papers*, vol. 39, no. 1, pp. 1–4, 2008.

[40] Y.-H. Kim, J.-S. Heo, T.-H. Kim, S. Park, M.-H. Yoon, J. Kim, M. S. Oh, G.-R. Yi, Y.-Y. Noh, and S. K. Park, "Flexible metal-oxide devices made by room-temperature photochemical activation of sol–gel films," *Nature*, vol. 489, no. 7414, pp. 128–132, Sep. 2012.

[41] T. Shimoda and S. Inoue, "Surface free technology by laser annealing (SUFTLA)," in *IEEE International Electron Devices Meeting (IEDM)*, 1999, pp. 289–292.

[42] H. Lifka, C. Tanase, D. McCulloch, P. Van de Weijer, and I. French, "53.4: Ultra-Thin Flexible OLED Device," *SID Symposium Digest of Technical Papers*, vol. 38, no. 1, pp. 1599–1602, 2007.

[43] S. Inoue, S. Utsunomiya, T. Saeki, and T. Shimoda, "Surface-free technology by laser annealing (SUFTLA) and its application to poly-Si TFT-LCDs on plastic film with integrated drivers," *IEEE Transactions on Electron Devices*, vol. 49, no. 8, pp. 1353–1360, Aug. 2002.

[44] S. Utsunomiya, T. Kamakura, M. Kasuga, M. Kimura, W. Miyazawa, S. Inoue, and T. Shimoda, "21.3: Flexible Color AM-OLED Display Fabricated Using Surface Free Technology by Laser Ablation/Annealing (SUFTLA®) and Ink-jet Printing Technology," *SID Symposium Digest of Technical Papers*, vol. 34, no. 1, pp. 864–867, 2003.

[45] N. Karaki, T. Nanmoto, H. Ebihara, S. Utsunomiya, S. Inoue, and T. Shimoda, "A flexible 8b asynchronous microprocessor based on low-temperature poly-silicon TFT technology," in *IEEE International Solid-State Circuits Conference (ISSCC)*, 2005, pp. 272–598.

[46] J.-S. Park, T.-W. Kim, D. Stryakhilev, J.-S. Lee, S.-G. An, Y.-S. Pyo, D.-B. Lee, Y. G. Mo, D.-U. Jin, and H. K. Chung, "Flexible full color organic light-emitting diode display on polyimide plastic substrate driven by amorphous indium gallium zinc oxide thin-film transistors," *Applied Physics Letters*, vol. 95, no. 1, pp. 013503–013503–3, Jul. 2009.

[47] M. Noda, N. Kobayashi, M. Katsuhara, A. Yumoto, S. Ushikura, R. Yasuda, N. Hirai, G. Yukawa, I. Yagi, K. Nomoto, and T. Urabe, "A Rollable AM-OLED Display Driven by OTFTs," *SID Symposium Digest of Technical Papers*, vol. 41, no. 1, pp. 710–713, 2010.

[48] W. Lim, E. A. Douglas, S.-H. Kim, D. P. Norton, S. J. Pearton, F. Ren, H. Shen, and W. H. Chang, "High mobility InGaZnO4 thin-film transistors on paper," *Applied Physics Letters*, vol. 94, no. 7, pp. 072103–072103–3, Feb. 2009.

[49] R. Martins, A. Nathan, R. Barros, L. Pereira, P. Barquinha, N. Correia, R. Costa, A. Ahnood, I. Ferreira, and E. Fortunato, "Complementary Metal Oxide Semiconductor Technology With and On Paper," *Advanced Materials*, vol. 23, no. 39, pp. 4491–4496, 2011.

[50] F. Eder, H. Klauk, M. Halik, U. Zschieschang, G. Schmid, and C. Dehm, "Organic electronics on paper," *Applied Physics Letters*, vol. 84, no. 14, pp. 2673–2675, Apr. 2004.

[51] U. Zschieschang, T. Yamamoto, K. Takimiya, H. Kuwabara, M. Ikeda, T. Sekitani, T. Someya, and H. Klauk, "Organic Electronics on Banknotes," *Advanced Materials*, vol. 23, no. 5, pp. 654–658, 2011.

[52] S. D. Theiss and S. Wagner, "Amorphous silicon thin-film transistors on steel foil substrates," *IEEE Electron Device Letters*, vol. 17, no. 12, pp. 578–580, Dec. 1996.

[53] T. Serikawa and F. Omata, "High-mobility poly-Si TFTs fabricated on flexible stainless-steel substrates," *IEEE Electron Device Letters*, vol. 20, no. 11, pp. 574–576, Nov. 1999.

[54] H. Gleskova and S. Wagner, "DC-gate-bias stressing of a-Si:H TFTs fabricated at 150 deg C on polyimide foil," *IEEE Transactions on Electron Devices*, vol. 48, no. 8, pp. 1667–1671, Aug. 2001.

[55] M. H. Lee, K.-Y. Ho, P.-C. Chen, C.-C. Cheng, S. T. Chang, M. Tang, M. H. Liao, and Y.-H. Yeh, "Promising a-Si:H TFTs with High Mechanical Reliability for Flexible Display," in *IEEE International Electron Devices Meeting (IEDM)*, 2006, pp. 1–4.

[56] P.-C. Yang, H.-Y. Chang, C.-H. Yang, C.-Y. Hsueh, H.-W. Lin, C.-Y. Chang, and S.-C. Lee, "Low Temperature Polycrystalline Silicon TFTs on Polyimide and Glass Substrates," in *IEEE Conference on Electron Devices and Solid-State Circuits (EDSSC)*, 2007, pp. 519–522.

[57] K. R. Sarma, "Active-matrix OLED using 150°C a-Si TFT backplane built on flexible plastic substrate," in *Annual Laser and Electro-Optics Society (LEOS) Meeting*, 2003, vol. 5080, pp. 180–191.

[58] R. A. Lujan and R. A. Street, "Flexible X-ray detector array fabricated with oxide thin-film transistors," *IEEE Electron Device Letters*, vol. 33, no. 5, pp. 688–690, May 2012.

[59] J. Fischer, M. Tietke, F. Fritze, O. Muth, M. Paeschke, D. Han, J. Kwack, T. Kim, J. Lee, S. Kim, and H. Chung, "Electronic passports with AMOLED displays," *Journal of the Society for Information Display*, vol. 19, no. 2, pp. 163–169, 2011.

[60] T. Someya, T. Sekitani, S. Iba, Y. Kato, H. Kawaguchi, and T. Sakurai, "A large-area, flexible pressure sensor matrix with organic field-effect transistors for artificial skin applications," *Proceedings of the National Academy of Sciences of the United States of America*, vol. 101, no. 27, pp. 9966–9970, Jul. 2004.

[61] H. Marien, M. S. J. Steyaert, E. van Veenendaal, and P. Heremans, "A fully integrated $\Delta\Sigma$ ADC in organic thin-film transistor technology on flexible plastic foil," *IEEE Journal of Solid-State Circuits*, vol. 46, no. 1, pp. 276–284, Jan. 2011.

[62] W. Xiong, Y. Guo, U. Zschieschang, H. Klauk, and B. Murmann, "A 3-V, 6-bit C-2C digital-to-analog converter using complementary organic thin-film transistors on glass," *IEEE Journal of Solid-State Circuits*, vol. 45, no. 7, pp. 1380–1388, Jul. 2010.

[63] K. Myny, E. van Veenendaal, G. H. Gelinck, J. Genoe, W. Dehaene, and P. Heremans, "An 8-bit, 40-instructions-per-second organic microprocessor on plastic foil," *IEEE Journal of Solid-State Circuits*, vol. 47, no. 1, pp. 284–291, Jan. 2012.

[64] H. Marien, M. Steyaert, N. van Aerle, and P. Heremans, "An analog organic first-order CT $\Delta\Sigma$ ADC on a flexible plastic substrate with 26.5dB precision," in *IEEE International Solid-State Circuits Conference (ISSCC)*, 2010, pp. 136–137.

[65] N. Morosawa, Y. Ohshima, M. Morooka, T. Arai, and T. Sasaoka, "Novel self-aligned top-gate oxide TFT for AMOLED displays," *Journal of the Society for Information Display*, vol. 20, no. 1, p. 47, 2012.

[66] E. J. Meijer, D. M. de Leeuw, S. Setayesh, E. van Veenendaal, B.-H. Huisman, P. W. M. Blom, J. C. Hummelen, U. Scherf, and T. M. Klapwijk, "Solution-processed ambipolar organic field-effect transistors and inverters," *Nature Materials*, vol. 2, no. 10, pp. 678–682, 2003.

[67] J. Zaumseil and H. Sirringhaus, "Electron and ambipolar transport in organic field-effect transistors," *Chemical Reviews*, vol. 107, no. 4, pp. 1296–323, 2007.

[68] H. Klauk, G. Schmid, W. Radlik, W. Weber, L. Zhou, C. D. Sheraw, J. A. Nichols, and T. N. Jackson, "Contact resistance in organic thin film transistors," *Solid-State Electronics*, vol. 47, no. 2, pp. 297–301, Feb. 2003.

[69] P. V. Necliudov, M. S. Shur, D. J. Gundlach, and T. N. Jackson, "Contact resistance extraction in pentacene thin film transistors," *Solid-State Electronics*, vol. 47, no. 2, pp. 259–262, Feb. 2003.

[70] O. Marinov, M. J. Deen, U. Zschieschang, and H. Klauk, "Organic thin-film transistors: Part I-compact DC modeling," *IEEE Transactions on Electron Devices*, vol. 56, no. 12, pp. 2952–2961, Dec. 2009.

[71] M. J. Deen, O. Marinov, U. Zschieschang, and H. Klauk, "Organic thin-film transistors: Part II: Parameter extraction," *IEEE Transactions on Electron Devices*, vol. 56, no. 12, pp. 2962–2968, Dec. 2009.

[72] F. Torricelli, "Charge transport in organic transistors accounting for a wide distribution of carrier energies: Part I: Theory," *IEEE Transactions on Electron Devices*, vol. 59, no. 5, pp. 1514–1519, May 2012.

[73] F. Torricelli, K. O'Neill, G. H. Gelinck, K. Myny, J. Genoe, and E. Cantatore, "Charge transport in organic transistors accounting for a wide distribution of carrier

energies: Part II: TFT modeling," *IEEE Transactions on Electron Devices*, vol. 59, no. 5, pp. 1520–1528, May 2012.

[74] L. Li, M. Debucquoy, J. Genoe, and P. Heremans, "A compact model for polycrystalline pentacene thin-film transistor," *Journal of Applied Physics*, vol. 107, no. 2, pp. 024519–024519-3, Jan. 2010.

[75] L. Li, H. Marien, J. Genoe, M. Steyaert, and P. Heremans, "Compact model for organic thin-film transistor," *IEEE Electron Device Letters*, vol. 31, no. 3, pp. 210–212, Mar. 2010.

[76] F. Torricelli, J. R. Meijboom, E. Smits, A. K. Tripathi, M. Ferroni, S. Federici, G. H. Gelinck, L. Colalongo, Z. M. Kovacs-Vajna, D. de Leeuw, and E. Cantatore, "Transport physics and device modeling of zinc oxide thin-film transistors: Part I: Long-channel devices," *IEEE Transactions on Electron Devices*, vol. 58, no. 8, pp. 2610–2619, Aug. 2011.

[77] F. Torricelli, E. C. P. Smits, J. R. Meijboom, A. K. Tripathi, G. H. Gelinck, L. Colalongo, Z. M. Kovacs-Vajna, D. M. de Leeuw, and E. Cantatore, "Transport Physics and Device Modeling of Zinc Oxide Thin-Film Transistors: Part II: Contact resistance in short channel devices," *IEEE Transactions on Electron Devices*, vol. 58, no. 9, pp. 3025–3033, Sep. 2011.

[78] H. He and X. Zheng, "Analytical drain current model for amorphous IGZO thin-film transistors in above-threshold regime," *Journal of Semiconductors*, vol. 32, no. 7, p. 074004, Jul. 2011.

[79] K. Abe, N. Kaji, H. Kumomi, K. Nomura, T. Kamiya, M. Hirano, and H. Hosono, "Simple analytical model of on operation of amorphous in-Ga-Zn-O thin-film transistors," *IEEE Transactions on Electron Devices*, vol. 58, no. 10, pp. 3463–3471, Oct. 2011.

[80] D. H. Kim, Y. W. Jeon, S. Kim, Y. Kim, Y. S. Yu, D. M. Kim, and H.-I. Kwon, "Physical parameter-based SPICE models for InGaZnO thin-film transistors applicable to process optimization and robust circuit design," *IEEE Electron Device Letters*, vol. 33, no. 1, pp. 59–61, Jan. 2012.

[81] M. R. M. Pherson, "The adjustment of MOS transistor threshold voltage by ion implantation," *Applied Physics Letters*, vol. 18, no. 11, pp. 502–504, Jun. 1971.

[82] T. Mizuno, J. Okumtura, and A. Toriumi, "Experimental study of threshold voltage fluctuation due to statistical variation of channel dopant number in MOSFET's," *IEEE Transactions on Electron Devices*, vol. 41, no. 11, pp. 2216–2221, Nov. 1994.

[83] I. Nausieda, K. K. Ryu, D. D. He, A. I. Akinwande, V. Bulovic, and C. G. Sodini, "Dual threshold voltage organic thin-film transistor technology," *IEEE Transactions on Electron Devices*, vol. 57, no. 11, pp. 3027–3032, Nov. 2010.

[84] S.-T. Han, Y. Zhou, Z.-X. Xu, and V. A. L. Roy, "Controllable threshold voltage shifts of polymer transistors and inverters by utilizing gold nanoparticles," *APL: Organic Electronics and Photonics*, vol. 5, no. 7, pp. 154–154, Jul. 2012.

[85] T. Yokota, T. Nakagawa, T. Sekitani, Y. Noguchi, K. Fukuda, U. Zschieschang, H. Klauk, K. Takeuchi, M. Takamiya, T. Sakurai, and T. Someya, "Control of threshold voltage in low-voltage organic complementary inverter circuits with floating gate structures," *Applied Physics Letters*, vol. 98, no. 19, pp. 193302–193302-3, May 2011.

[86] S. Kobayashi, T. Nishikawa, T. Takenobu, S. Mori, T. Shimoda, T. Mitani, H. Shimotani, N. Yoshimoto, S. Ogawa, and Y. Iwasa, "Control of carrier density by self-assembled monolayers in organic field-effect transistors," *Nature Materials*, vol. 3, no. 5, pp. 317–322, 2004.

[87] K. P. Pernstich, S. Haas, D. Oberhoff, C. Goldmann, D. J. Gundlach, B. Batlogg, A. N. Rashid, and G. Schitter, "Threshold voltage shift in organic field effect transistors by dipole monolayers on the gate insulator," *Journal of Applied Physics*, vol. 96, no. 11, pp. 6431–6438, Dec. 2004.

[88] M. Kitamura, Y. Kuzumoto, S. Aomori, M. Kamura, J. H. Na, and Y. Arakawa, "Threshold voltage control of bottom-contact n-channel organic thin-film transistors using modified drain/source electrodes," *Applied Physics Letters*, vol. 94, no. 8, pp. 083310–083310-3, Feb. 2009.

[89] G. H. Gelinck, E. van Veenendaal, and R. Coehoorn, "Dual-gate organic thin-film transistors," *Applied Physics Letters*, vol. 87, no. 7, pp. 073508–073508-3, Aug. 2005.

[90] S. Iba, T. Sekitani, Y. Kato, T. Someya, H. Kawaguchi, M. Takamiya, T. Sakurai, and S. Takagi, "Control of threshold voltage of organic field-effect transistors with double-gate structures," *Applied Physics Letters*, vol. 87, no. 2, pp. 023509–023509-3, Jul. 2005.

[91] M. Morana, G. Bret, and C. Brabec, "Double-gate organic field-effect transistor," *Applied Physics Letters*, vol. 87, no. 15, pp. 153511–153511-3, Oct. 2005.

[92] M.-J. Spijkman, K. Myny, E. C. P. Smits, P. Heremans, P. W. M. Blom, and D. M. de Leeuw, "Dual-gate thin-film transistors, integrated circuits and sensors," *Advanced Materials*, vol. 23, no. 29, pp. 3231–3242, 2011.

[93] J. J. Brondijk, M. Spijkman, F. Torricelli, P. W. M. Blom, and D. M. de Leeuw, "Charge transport in dual-gate organic field-effect transistors," *Applied Physics Letters*, vol. 100, no. 2, pp. 023308–023308-4, Jan. 2012.

[94] G. H. Gelinck, H. E. A. Huitema, E. van Veenendaal, E. Cantatore, L. Schrijnemakers, J. B. P. H. van der Putten, T. C. T. Geuns, M. Beenhakkers, J. B. Giesbers, B.-H. Huisman, E. J. Meijer, E. M. Benito, F. J. Touwslager, A. W. Marsman, B. J. E. van Rens, and D. M. de Leeuw, "Flexible active-matrix displays and shift registers based on solution-processed organic transistors," *Nature Materials*, vol. 3, no. 2, pp. 106–110, Jan. 2004.

[95] H. E. A. Huitema, "Rollable displays: The start of a new mobile device generation," presented at the Proc. 7th Annu. USDC Flexible Electron. Displays Conf., Phoenix, AZ, 2008.

[96] C. R. Kagan, A. Afzali, and T. O. Graham, "Operational and environmental stability of pentacene thin-film transistors," *Applied Physics Letters*, vol. 86, no. 19, pp. 193505–193505-3, May 2005.

[97] K. Myny, S. Steudel, P. Vicca, M. J. Beenhakkers, N. A. J. M. van Aerle, G. H. Gelinck, J. Genoe, W. Dehaene, and P. Heremans, "Plastic circuits and tags for 13.56 MHz radio-frequency communication," *Solid-State Electronics*, vol. 53, no. 12, pp. 1220–1226, Dec. 2009.

[98] S. De Vusser, S. Steudel, K. Myny, J. Genoe, and P. Heremans, "Integrated shadow mask method for patterning small molecule organic semiconductors," *Applied Physics Letters*, vol. 88, no. 10, pp. 103501–103501-3, Mar. 2006.

[99] A. K. Tripathi, E. C. P. Smits, J. B. P. H. van der Putten, M. van Neer, K. Myny, M. Nag, S. Steudel, P. Vicca, K. O'Neill, E. van Veenendaal, G. Genoe, P. Heremans, and G. H. Gelinck, "Low-voltage gallium–indium–zinc–oxide thin film transistors based logic circuits on thin plastic foil: Building blocks for radio frequency identification application," *Applied Physics Letters*, vol. 98, p. 162102, 2011.

[100] M. Rockelé, D.-V. Pham, A. Hoppe, J. Steiger, S. Botnaras, M. Nag, S. Steudel, K. Myny, S. Schols, R. Müller, B. van der Putten, J. Genoe, and P. Heremans,

"Low-temperature and scalable complementary thin-film technology based on solution-processed metal oxide n-TFTs and pentacene p-TFTs," *Organic Electronics*, vol. 12, no. 11, pp. 1909–1913, Nov. 2011.

[101] K. Myny, M. Rockele, A. Chasin, D. Pham, J. Steiger, S. Botnaras, D. Weber, B. Herold, J. Ficker, B. van der Putten, G. Gelinck, J. Genoe, W. Dehaene, and P. Heremans, "Bidirectional communication in an HF hybrid organic/solution-processed metal-oxide RFID tag," in *IEEE International Solid-State Circuits Conference (ISSCC)*, 2012, pp. 312–314.

[102] M. Rockelé, D.-V. Pham, J. Steiger, S. Botnaras, D. Weber, J. Vanfleteren, T. Sterken, D. Cuypers, S. Steudel, K. Myny, S. Schols, B. van der Putten, J. Genoe, and P. Heremans, "Solution-processed and low-temperature metal oxide n-channel thin-film transistors and low-voltage complementary circuitry on large-area flexible polyimide foil," *Journal of the Society for Information Display*, vol. 20, no. 9, pp. 499–507, 2012.

[103] B. Crone, A. Dodabalapur, Y.-Y. Lin, R. W. Filas, Z. Bao, A. LaDuca, R. Sarpeshkar, H. E. Katz, and W. Li, "Large-scale complementary integrated circuits based on organic transistors," *Nature*, vol. 403, no. 6769, pp. 521–523, Feb. 2000.

[104] P. van Lieshout, E. van Veenendaal, L. Schrijnemakers, G. Gelinck, F. Touwslager, and E. Huitema, "A flexible 240×320-pixel display with integrated row drivers manufactured in organic electronics," in *IEEE International Solid-State Circuits Conference (ISSCC)*, 2005, Vol. 1, pp. 578–618.

[105] M. Noda, N. Kobayashi, M. Katsuhara, A. Yumoto, S. Ushikura, R. Yasuda, N. Hirai, G. Yukawa, I. Yagi, K. Nomoto, and T. Urabe, "An OTFT-driven rollable OLED display," *Journal of the Society for Information Display*, vol. 19, no. 4, pp. 316–322, 2011.

[106] E. Cantatore, T. C. T. Geuns, A. F. A. Gruijthuijsen, G. H. Gelinck, S. Drews, and D. M. de Leeuw, "A 13.56MHz RFID system based on organic transponders," in *IEEE International Solid-State Circuits Conference (ISSCC)*, 2006, pp. 1042–1051.

[107] E. Cantatore, T. C. T. Geuns, G. H. Gelinck, E. van Veenendaal, A. F. A. Gruijthuijsen, L. Schrijnemakers, S. Drews, and D. M. de Leeuw, "A 13.56-MHz RFID system based on organic transponders," *IEEE Journal of Solid-State Circuits*, vol. 42, no. 1, pp. 84–92, Jan. 2007.

[108] A. Ullmann, M. Bohm, J. Krumm, and W. Fix, "Polymer multi-bit RFID transponder," in *International Conference on Organic Electronics (ICOE)*, Eindhoven, The Netherlands, 2007.

[109] K. Myny, S. Van Winckel, S. Steudel, P. Vicca, S. De Jonge, M. J. Beenhakkers, C. W. Sele, N. A. J. M. van Aerle, G. H. Gelinck, J. Genoe, and P. Heremans, "An inductively-coupled 64b organic RFID tag operating at 13.56MHz with a data rate of 787b/s," in *IEEE International Solid-State Circuits Conference (ISSCC)*, 2008, pp. 290–291.

[110] K. Myny, M. J. Beenhakkers, N. A. J. M. van Aerle, G. H. Gelinck, J. Genoe, W. Dehaene, and P. Heremans, "A 128b organic RFID transponder chip, including Manchester encoding and ALOHA anti-collision protocol, operating with a data rate of 1529b/s," in *IEEE International Solid-State Circuits Conference (ISSCC)*, 2009, pp. 206–207.

[111] J. B. Koo, J. W. Lim, S. H. Kim, C. H. Ku, S. C. Lim, J. H. Lee, S. J. Yun, and Y. S. Yang, "Pentacene thin-film transistors and inverters with dual-gate structure," *Electrochemical and Solid-State Letters*, vol. 9, no. 11, p. G320, 2006.

[112] J. B. Koo, C. H. Ku, J. W. Lim, and S. H. Kim, "Novel organic inverters with dual-gate pentacene thin-film transistor," *Organic Electronics*, vol. 8, no. 5, pp. 552–558, Oct. 2007.

[113] M. Spijkman, E. C. P. Smits, P. W. M. Blom, D. M. de Leeuw, Y. Bon Saint Côme, S. Setayesh, and E. Cantatore, "Increasing the noise margin in organic circuits using dual gate field-effect transistors," *Applied Physics Letters*, vol. 92, no. 14, pp. 143304–143304-3, Apr. 2008.

[114] K. Myny, M. J. Beenhakkers, N. A. J. van Aerle, G. H. Gelinck, J. Genoe, W. Dehaene, and P. Heremans, "Robust digital design in organic electronics by dual-gate technology," in *IEEE International Solid-State Circuits Conference (ISSCC)*, 2010, pp. 140–141.

[115] K. Myny, M. J. Beenhakkers, N. A. J. van Aerle, G. H. Gelinck, J. Genoe, W. Dehaene, and P. Heremans, "Unipolar organic transistor circuits made robust by dual-gate technology," *IEEE Journal of Solid-State Circuits*, vol. 46, no. 5, pp. 1223–1230, May 2011.

[116] K. Myny, E. van Veenendaal, G. H. Gelinck, J. Genoe, W. Dehaene, and P. Heremans, "An 8b organic microprocessor on plastic foil," in *IEEE International Solid-State Circuits Conference (ISSCC)*, 2011, pp. 322–324.

[117] R. Blache, J. Krumm, and W. Fix, "Organic CMOS circuits for RFID applications," in *Solid-State Circuits Conference – Digest of Technical Papers, 2009. ISSCC 2009. IEEE International*, 2009, pp. 208–209.

[118] K. Myny, S. Steudel, S. Smout, P. Vicca, F. Furthner, B. van der Putten, A. K. Tripathi, G. H. Gelinck, J. Genoe, W. Dehaene, and P. Heremans, "Organic RFID transponder chip with data rate compatible with electronic product coding," *Organic Electronics*, vol. 11, no. 7, pp. 1176–1179, Jul. 2010.

[119] H. Klauk, U. Zschieschang, J. Pflaum, and M. Halik, "Ultralow-power organic complementary circuits," *Nature*, vol. 445, no. 7129, pp. 745–748, Feb. 2007.

[120] K. Ishida, N. Masunaga, Z. Zhou, T. Yasufuku, T. Sekitani, U. Zschieschang, H. Klauk, M. Takamiya, T. Someya, and T. Sakurai, "Stretchable EMI measurement sheet with 8 × 8 coil array, 2 V organic CMOS decoder, and 0.18μm silicon CMOS LSIs for electric and magnetic field detection," *IEEE Journal of Solid-State Circuits*, vol. 45, no. 1, pp. 249–259, Jan. 2010.

[121] K. Ishida, N. Masunaga, R. Takahashi, T. Sekitani, S. Shino, U. Zschieschang, H. Klauk, M. Takamiya, T. Someya, and T. Sakurai, "User customizable logic paper (UCLP) with sea-of transmission-gates (SOTG) of 2-V organic CMOS and ink-jet printed interconnects," *IEEE Journal of Solid-State Circuits*, vol. 46, no. 1, pp. 285–292, Jan. 2011.

[122] K. Fukuda, T. Sekitani, U. Zschieschang, H. Klauk, K. Kuribara, T. Yokota, T. Sugino, K. Asaka, M. Ikeda, H. Kuwabara, T. Yamamoto, K. Takimiya, T. Fukushima, T. Aida, M. Takamiya, T. Sakurai, and T. Someya, "A 4 V operation, flexible braille display using organic transistors, carbon nanotube actuators, and organic static random-access memory," *Advanced Functional Materials*, vol. 21, no. 21, pp. 4019–4027, 2011.

[123] W. Xiong, U. Zschieschang, H. Klauk, and B. Murmann, "A 3V 6b successive-approximation ADC using complementary organic thin-film transistors on glass," in *IEEE International Solid-State Circuits Conference (ISSCC)*, 2010, pp. 134–135.

[124] H. Klauk, M. Halik, U. Zschieschang, F. Eder, G. Schmid, and C. Dehm, "Pentacene organic transistors and ring oscillators on glass and on flexible polymeric substrates," *Applied Physics Letters*, vol. 82, no. 23, pp. 4175–4177, Jun. 2003.

[125] K. Fukuda, T. Sekitani, T. Yokota, K. Kuribara, T.-C. Huang, T. Sakurai, U. Zschieschang, H. Klauk, M. Ikeda, H. Kuwabara, T. Yamamoto, K. Takimiya, K.-T. Cheng, and T. Someya, "Organic pseudo-CMOS circuits for low-voltage large-gain high-speed operation," *IEEE Electron Device Letters*, vol. 32, no. 10, pp. 1448–1450, Oct. 2011.

[126] H. Marien, M. Steyaert, S. Steudel, P. Vicca, S. Smout, G. Gelinck, and P. Heremans, "An organic integrated capacitive DC-DC up-converter," in *European Solid-State Circuits Conference (ESSCIRC)*, 2010, pp. 510–513.

[127] H. Marien, M. Steyaert, E. van Veenendaal, and P. Heremans, "Organic dual DC-DC upconverter on foil for improved circuit reliability," *Electronics Letters*, vol. 47, no. 4, pp. 278–280, 2011.

[128] H. Marien, M. Steyaert, E. van Veenendaal, and P. Heremans, "DC-DC converter assisted two-stage amplifier in organic thin-film transistor technology on foil," in *European Solid-State Circuits Conference (ESSCIRC)*, 2011, pp. 411–414.

[129] Y.-S. Park, D.-Y. Kim, K.-N. Kim, Y. Matsueda, J.-H. Choi, C.-K. Kang, H.-D. Kim, H. K. Chung, and O.-K. Kwon, "An 8b source driver for 2.0 inch full-color active-matrix OLEDs made with LTPS TFTs," in *IEEE International Solid-State Circuits Conference (ISSCC)*, 2007, pp. 130–592.

[130] T. Zaki, F. Ante, U. Zschieschang, J. Butschke, F. Letzkus, H. Richter, H. Klauk, and J. N. Burghartz, "A 3.3 V 6-bit 100 kS/s current-steering digital-to-analog converter using organic P-type thin-film transistors on glass," *IEEE Journal of Solid-State Circuits*, vol. 47, no. 1, pp. 292–300, Jan. 2012.

[131] D. Raiteri, F. Torricelli, K. Myny, M. Nag, B. Van der Putten, E. Smits, S. Steudel, K. Tempelaars, A. Tripathi, G. Gelinck, A. Van Roermund, and E. Cantatore, "A 6b 10MS/s current-steering DAC manufactured with amorphous Gallium-Indium-Zinc-Oxide TFTs achieving SFDR > 30dB up to 300kHz," in *IEEE International Solid-State Circuits Conference (ISSCC)*, 2012, pp. 314–316.

[132] J. M. Rabaey, A. Chandrakasan, and B. Nikolic, *Digital Integrated Circuits*, 2nd ed. Prentice Hall, 2003.

[133] C. F. Hill, "Noise margin and noise immunity in logic circuits," *Microelectronics*, vol. 1, pp. 16–21, Apr. 1968.

[134] E. Seevinck, F. J. List, and J. Lohstroh, "Static-noise margin analysis of MOS SRAM cells," *IEEE Journal of Solid-State Circuits*, vol. 22, no. 5, pp. 748–754, Oct. 1987.

[135] J. R. Hauser, "Noise margin criteria for digital logic circuits," *IEEE Transactions on Education*, vol. 36, no. 4, pp. 363–368, Nov. 1993.

[136] J. Lohstroh, E. Seevinck, and J. de Groot, "Worst-case static noise margin criteria for logic circuits and their mathematical equivalence," *IEEE Journal of Solid-State Circuits*, vol. 18, no. 6, pp. 803–807, Dec. 1983.

[137] S. De Vusser, J. Genoe, and P. Heremans, "Influence of transistor parameters on the noise margin of organic digital circuits," *IEEE Transactions on Electron Devices*, vol. 53, no. 4, pp. 601–610, Apr. 2006.

[138] H. Sirringhaus, T. Kawase, R. H. Friend, T. Shimoda, M. Inbasekaran, W. Wu, and E. P. Woo, "High-Resolution Inkjet Printing of All-Polymer Transistor Circuits," *Science*, vol. 290, no. 5499, pp. 2123–2126, Dec. 2000.

[139] M. Halik, H. Klauk, U. Zschieschang, G. Schmid, C. Dehm, M. Schütz, S. Maisch, F. Effenberger, M. Brunnbauer, and F. Stellacci, "Low-voltage organic transistors with an amorphous molecular gate dielectric," *Nature*, vol. 431, no. 7011, pp. 963–966, Oct. 2004.

[140] K. Kim and Y. Kim, "Intrinsic capacitance characteristics of top-contact organic thin-film transistors," *IEEE Transactions on Electron Devices*, vol. 57, no. 9, pp. 2344–2347, Sep. 2010.

[141] F. Torricelli, Z. M. Kovacs-Vajna, and L. Colalongo, "A charge-based OTFT model for circuit simulation," *IEEE Transactions on Electron Devices*, vol. 56, no. 1, pp. 20–30, Jan. 2009.

[142] H. Klauk, D. J. Gundlach, and T. N. Jackson, "Fast organic thin-film transistor circuits," *IEEE Electron Device Letters*, vol. 20, no. 6, pp. 289–291, Jun. 1999.

[143] S. Borkar, T. Karnik, S. Narendra, J. Tschanz, A. Keshavarzi, and V. De, "Parameter variations and impact on circuits and microarchitecture," in *Design Automation Conference (DAC)*, 2003, pp. 338–342.

[144] A. Asenov, A. R. Brown, J. H. Davies, S. Kaya, and G. Slavcheva, "Simulation of intrinsic parameter fluctuations in decananometer and nanometer-scale MOSFETs," *IEEE Transactions on Electron Devices*, vol. 50, no. 9, pp. 1837–1852, Sep. 2003.

[145] K. Bernstein, D. J. Frank, A. E. Gattiker, W. Haensch, B. L. Ji, S. R. Nassif, E. J. Nowak, D. J. Pearson, and N. J. Rohrer, "High-performance CMOS variability in the 65-nm regime and beyond," *IBM Journal of Research and Development*, vol. 50, no. 4/5, pp. 433–449, Jul. 2006.

[146] K. J. Kuhn, "Reducing Variation in Advanced Logic Technologies: Approaches to Process and Design for Manufacturability of Nanoscale CMOS," in *IEEE International Electron Devices Meeting (IEDM)*, 2007, pp. 471–474.

[147] D. Sylvester, K. Agarwal, and S. Shah, "Variability in nanometer CMOS: Impact, analysis, and minimization," *Integration, the VLSI Journal*, vol. 41, no. 3, pp. 319–339, May 2008.

[148] H. Soeleman, K. Roy, and B. C. Paul, "Robust subthreshold logic for ultra-low power operation," *IEEE Transactions on Very Large Scale Integration (VLSI) Systems*, vol. 9, no. 1, pp. 90–99, Feb. 2001.

[149] J.-J. Kim and K. Roy, "Double gate-MOSFET subthreshold circuit for ultralow power applications," *IEEE Transactions on Electron Devices*, vol. 51, no. 9, pp. 1468–1474, Sep. 2004.

[150] J. Kwong and A. P. Chandrakasan, "Variation-driven device sizing for minimum energy sub-threshold circuits," in *International Symposium on Low Power Electronics and Design (ISLPED)*, 2006, pp. 8–13.

[151] S. Fisher, A. Teman, D. Vaysman, A. Gertsman, O. Yadid-Pecht, and A. Fish, "Digital subthreshold logic design – motivation and challenges," in *IEEE 25th Convention of Electrical and Electronics Engineers in Israel (IEEEI)*, 2008, pp. 702–706.

[152] S. Verlaak, S. Steudel, P. Heremans, D. Janssen, and M. Deleuze, "Nucleation of organic semiconductors on inert substrates," *Physical Review B*, vol. 68, no. 19, Nov. 2003.

[153] S. Verlaak, V. Arkhipov, and P. Heremans, "Modeling of transport in polycrystalline organic semiconductor films," *Applied Physics Letters*, vol. 82, no. 5, pp. 745–747, Feb. 2003.

[154] X. Li, A. Kadashchuk, I. I. Fishchuk, W. T. T. Smaal, G. Gelinck, D. J. Broer, J. Genoe, P. Heremans, and H. Bässler, "Electric field confinement effect on charge

transport in organic field-effect transistors," *Physical Review Letters*, vol. 108, no. 6, p. 066601, Feb. 2012.

[155] S. Verlaak and P. Heremans, "Molecular microelectrostatic view on electronic states near pentacene grain boundaries," *Physical Review B*, vol. 75, no. 11, p. 115127, Mar. 2007.

[156] X. Li, W. T. T. Smaal, C. Kjellander, B. van der Putten, K. Gualandris, E. C. P. Smits, J. Anthony, D. J. Broer, P. W. M. Blom, J. Genoe, and G. Gelinck, "Charge transport in high-performance ink-jet printed single-droplet organic transistors based on a silylethynyl substituted pentacene/insulating polymer blend," *Organic Electronics*, vol. 12, no. 8, pp. 1319–1327, Aug. 2011.

[157] S. Steudel, S. De Vusser, S. De Jonge, D. Janssen, S. Verlaak, J. Genoe, and P. Heremans, "Influence of the dielectric roughness on the performance of pentacene transistors," *Applied Physics Letters*, vol. 85, no. 19, pp. 4400–4402, Nov. 2004.

[158] A. Mityashin, Y. Olivier, T. Van Regemorter, C. Rolin, S. Verlaak, N. G. Martinelli, D. Beljonne, J. Cornil, J. Genoe, and P. Heremans, "Unraveling the mechanism of molecular doping in organic semiconductors," *Advanced Materials*, vol. 24, no. 12, pp. 1535–1539, 2012.

[159] M. Debucquoy, S. Verlaak, S. Steudel, K. Myny, J. Genoe, and P. Heremans, "Correlation between bias stress instability and phototransistor operation of pentacene thin-film transistors," *Applied Physics Letters*, vol. 91, no. 10, pp. 103508–103508-3, Sep. 2007.

[160] C.-Y. Chen, S.-D. Wang, M.-S. Shieh, W.-C. Chen, H.-Y. Lin, K.-L. Yeh, J.-W. Lee, and T.-F. Lei, "Plasma-Induced Damage on the Performance and Reliability of Low-Temperature Polycrystalline Silicon Thin-Film Transistors," *Journal of the Electrochemical Society*, vol. 154, no. 1, pp. H30–H35, Jan. 2007.

[161] K. Eriguchi, Y. Nakakubo, A. Matsuda, M. Kamei, Y. Takao, and K. Ono, "Comprehensive modeling of threshold voltage variability induced by plasma damage in advanced metal–oxide–semiconductor field-effect transistors," *Japanese Journal of Applied Physics*, vol. 49, no. 4, p. 04DA18, 2010.

[162] B. Kim, S. H. Kwon, K. H. Kwon, K.-H. Baek, J. H. Lee, D. H. Kim, and G. S. May, "Statistical characterization of process-induced plasma damage," *Materials and Manufacturing Processes*, vol. 24, no. 6, pp. 610–614, 2009.

[163] M. Halik, H. Klauk, U. Zschieschang, T. Kriem, G. Schmid, W. Radlik, and K. Wussow, "Fully patterned all-organic thin film transistors," *Applied Physics Letters*, vol. 81, no. 2, pp. 289–291, Jul. 2002.

[164] H.-Y. Tseng and V. Subramanian, "All inkjet printed self-aligned transistors and circuits applications," in *IEEE International Electron Devices Meeting (IEDM)*, 2009, pp. 1–4.

[165] A. de la Fuente Vornbrock, D. Sung, H. Kang, R. Kitsomboonloha, and V. Subramanian, "Fully gravure and ink-jet printed high speed pBTTT organic thin film transistors," *Organic Electronics*, vol. 11, no. 12, pp. 2037–2044, Dec. 2010.

[166] I. G. Hill, "Numerical simulations of contact resistance in organic thin-film transistors," *Applied Physics Letters*, vol. 87, no. 16, pp. 163505–163505-3, Oct. 2005.

[167] P. Barquinha, A. M. Vila, G. Goncalves, L. Pereira, R. Martins, J. R. Morante, and E. Fortunato, "Gallium-indium-zinc-oxide-based thin-film transistors: Influence of the source/drain material," *IEEE Transactions on Electron Devices*, vol. 55, no. 4, pp. 954–960, Apr. 2008.

[168] Y. Shimura, K. Nomura, H. Yanagi, T. Kamiya, M. Hirano, and H. Hosono, "Specific contact resistances between amorphous oxide semiconductor In–Ga–Zn–O and metallic electrodes," *Thin Solid Films*, vol. 516, no. 17, pp. 5899–5902, Jul. 2008.

[169] W.-S. Kim, Y.-K. Moon, K.-T. Kim, J.-H. Lee, B. Ahn, and J.-W. Park, "An investigation of contact resistance between metal electrodes and amorphous gallium–indium–zinc oxide (a-GIZO) thin-film transistors," *Thin Solid Films*, vol. 518, no. 22, pp. 6357–6360, Sep. 2010.

[170] M. Weis, J. Lin, D. Taguchi, T. Manaka, and M. Iwamoto, "Insight into the contact resistance problem by direct probing of the potential drop in organic field-effect transistors," *Applied Physics Letters*, vol. 97, no. 26, pp. 263304–263304–3, Dec. 2010.

[171] M. Marinkovic, D. Belaineh, V. Wagner, and D. Knipp, "On the origin of contact resistances of organic thin film transistors," *Advanced Materials*, vol. 24, no. 29, pp. 4005–4009, 2012.

[172] I. Yakimets, D. MacKerron, P. Giesen, K. J. Kilmartin, M. Goorhuis, E. Meinders, and W. A. MacDonald, "Polymer Substrates for Flexible Electronics: Achievements and Challenges," *Advanced Materials Research*, vol. 93 –94, pp. 5–8, Jan. 2010.

[173] D. Bode, C. Rolin, S. Schols, M. Debucquoy, S. Steudel, G. H. Gelinck, J. Genoe, and P. Heremans, "Noise-margin analysis for organic thin-film complementary technology," *IEEE Transactions on Electron Devices*, vol. 57, no. 1, pp. 201–208, Jan. 2010.

[174] D. Bode, "Complementary Technology for Organic Thin-Film Transistors," PhD dissertation, KULeuven, Leuven, 2011.

[175] K. Myny, P. van Lieshout, J. Genoe, W. Dehaene, and P. Heremans, "Accounting for variability in the design of circuits with organic thin-film transistors," *Organic Electronics*, vol. 15, no. 4, pp. 937–942, Apr. 2014.

[176] M. J. M. Pelgrom, A. C. J. Duinmaijer, and A. P. G. Welbers, "Matching properties of MOS transistors," *IEEE Journal of Solid-State Circuits*, vol. 24, no. 5, pp. 1433–1439, Oct. 1989.

[177] J. Tschanz, J. Kao, S. Narendra, R. Nair, D. Antoniadis, A. Chandrakasan, and V. De, "Adaptive body bias for reducing impacts of die-to-die and within-die parameter variations on microprocessor frequency and leakage," in *IEEE International Solid-State Circuits Conference (ISSCC)*, 2002, vol. 1, pp. 422–478.

[178] Q. Liu and S. S. Sapatnekar, "Capturing post-silicon variations using a representative critical path," *IEEE Transactions on Computer-Aided Design of Integrated Circuits and Systems*, vol. 29, no. 2, pp. 211–222, Feb. 2010.

[179] G. Gelinck, P. Heremans, K. Nomoto, and T. D. Anthopoulos, "Organic transistors in optical displays and microelectronic applications," *Advanced Materials*, vol. 22, no. 34, pp. 3778–3798, 2010.

[180] T. Someya, Y. Kato, T. Sekitani, S. Iba, Y. Noguchi, Y. Murase, H. Kawaguchi, and T. Sakurai, "Conformable, flexible, large-area networks of pressure and thermal sensors with organic transistor active matrixes," *Proceedings of the National Academy of Sciences of the United States of America*, vol. 102, no. 35, pp. 12321–12325, Aug. 2005.

[181] T. Sekitani, U. Zschieschang, H. Klauk, and T. Someya, "Flexible organic transistors and circuits with extreme bending stability," *Nature Materials*, vol. 9, no. 12, pp. 1015–1022, 2010.

[182] S. K. Hong, "A Study on Compensation and Driving Circuits for AMOLED Display," PhD dissertation, Kyung Hee University, Seoul, Korea, 2010.

[183] C.-L. Fan, H.-L. Lai, and J.-Y. Chang, "Improvement in brightness uniformity by compensating for the threshold voltages of both the driving thin-film transistor and the organic light-emitting diode for active-matrix organic light-emitting diode displays," *Japanese Journal of Applied Physics*, vol. 49, no. 5, p. 05EB04, May 2010.

[184] N. D. Jankovic and V. Brajovic, "Vth compensated AMOLED pixel employing dual-gate TFT driver," *Electronics Letters*, vol. 47, no. 7, p. 456, 2011.

[185] K. Oh and O.-K. Kwon, "Threshold-voltage-shift compensation and suppression method using hydrogenated amorphous silicon thin-film transistors for large active matrix organic light-emitting diode displays," *Japanese Journal of Applied Physics*, vol. 51, p. 03CD01, Mar. 2012.

[186] Y.-H. Tai, L.-S. Chou, H.-L. Chiu, and B.-C. Chen, "Three-transistor AMOLED pixel circuit with threshold voltage compensation function using dual-gate IGZO TFT," *IEEE Electron Device Letters*, vol. 33, no. 3, pp. 393–395, Mar. 2012.

[187] M. Bohm, A. Ullmann, D. Zipperer, A. Knobloch, W. H. Glauert, and W. Fix, "Printable electronics for polymer RFID applications," in *IEEE International Solid-State Circuits Conference (ISSCC)*, 2006, pp. 1034–1041.

[188] M. Jung, J. Kim, J. Noh, N. Lim, C. Lim, G. Lee, J. Kim, H. Kang, K. Jung, A. D. Leonard, J. M. Tour, and G. Cho, "All-printed and roll-to-roll-printable 13.56-MHz-operated 1-bit RF tag on plastic foils," *IEEE Transactions on Electron Devices*, vol. 57, no. 3, pp. 571–580, Mar. 2010.

[189] G. Cho, "Roll-to-Roll Printed 13.56 MHz Operated 16-Bit RFID Tags and Smart RF Logos," in *Printed Electronics and Photovoltaics Europe*, Dresden, Germany, 2010.

[190] D. K. Finkenzeller, *RFID Handbook: Fundamentals and Applications in Contactless Smart Cards, Radio Frequency Identification and Near-Field Communication*, 3rd ed. John Wiley & Sons, 2010.

[191] A. W. Marsman, C. M. Hart, G. H. Gelinck, T. C. T. Geuns, and D. M. de Leeuw, "Doped polyaniline polymer fuses: Electrically programmable read-only-memory elements," *Journal of Materials Research*, vol. 19, no. 07, pp. 2057–2060, 2004.

[192] K. Myny, S. Steudel, P. Vicca, J. Genoe, and P. Heremans, "An integrated double half-wave organic Schottky diode rectifier on foil operating at 13.56 MHz," *Applied Physics Letters*, vol. 93, p. 093305, 2008.

[193] B. N. Pal, J. Sun, B. J. Jung, E. Choi, A. G. Andreou, and H. E. Katz, "Pentacene-zinc oxide vertical diode with compatible grains and 15-MHz rectification," *Advanced Materials*, vol. 20, no. 5, pp. 1023–1028, 2008.

[194] S. Steudel, K. Myny, V. Arkhipov, C. Deibel, S. De Vusser, J. Genoe, and P. Heremans, "50 MHz rectifier based on an organic diode," *Nature Materials*, vol. 4, no. 8, pp. 597–600, 2005.

[195] S. Steudel, S. De Vusser, K. Myny, M. Lenes, J. Genoe, and P. Heremans, "Comparison of organic diode structures regarding high-frequency rectification behavior in radio-frequency identification tags," *Journal of Applied Physics*, vol. 99, no. 11, pp. 114519–114519–7, Jun. 2006.

[196] A. Chasin, S. Steudel, K. Myny, M. Nag, T.-H. Ke, S. Schols, J. Genoe, G. Gielen, and P. Heremans, "High-performance a-In-Ga-Zn-O Schottky diode with oxygen-treated metal contacts," *Applied Physics Letters*, vol. 101, no. 11, pp. 113505–113505–5, Sep. 2012.

[197] T. Kawamura, H. Wakana, K. Fujii, H. Ozaki, K. Watanabe, T. Yamazoe, H. Uchiyama, and K. Torii, "Oxide TFT rectifier achieving 13.56-MHz wireless operation with DC

output up to 12 V," in *IEEE International Electron Devices Meeting (IEDM)*, 2010, pp. 21.4.1–21.4.4.

[198] "EPC standard" [Online]. Available: http://www.epcglobalinc.org/standards/specs/.

[199] H. Ozaki, T. Kawamura, H. Wakana, T. Yamazoe, and H. Uchiyama, "20-μW operation of an a-IGZO TFT-based RFID chip using purely NMOS 'active' load logic gates with ultra-low-consumption power," in *2011 Symposium on VLSI Circuits (VLSIC)*, 2011, pp. 54–55.

[200] K. Myny, M. Rockele, A. Chasin, D.-V. Pham, J. Steiger, S. Botnaras, D. Weber, B. Herold, J. Ficker, B. D. van Putten, G. H. Gelinck, J. Genoe, W. Dehaene, and P. Heremans, "Bidirectional Communication in an HF Hybrid Organic/ Solution-Processed Metal-Oxide RFID Tag," *IEEE Transactions on Electron Devices*, vol. 61, no. 7, pp. 2387–2393, Jul. 2014.

[201] K. Myny, S. Smout, M. Rockelé, A. Bhoolokam, T. H. Ke, S. Steudel, B. Cobb, A. Gulati, F. G. Rodriguez, K. Obata, M. Marinkovic, D.-V. Pham, A. Hoppe, G. H. Gelinck, J. Genoe, W. Dehaene, and P. Heremans, "A thin-film microprocessor with inkjet print-programmable memory," *Sci. Rep.*, vol. 4, Dec. 2014.

[202] J. A. Rogers, T. Someya, and Y. Huang, "Materials and Mechanics for Stretchable Electronics," *Science*, vol. 327, no. 5973, pp. 1603–1607, Mar. 2010.

[203] "Historic data are collected on the Intel Museum" [Online]. Available: http://www .intel.com/about/companyinfo/museum/exhibits/4004/index.htm. The specifications can be found at http://datasheets.chipdb.org/Intel/MCS-4/datashts/intel-4004.pdf

Index